Human growth and development

Veronica Windmill

Edward Arnold

First published in Great Britain 1987
by Edward Arnold (Publishers) Ltd
41 Bedford Square
London WC1B 3DQ

Edward Arnold (Australia) Pty Ltd
80 Waverley Road
Caulfield East
Victoria 3145
Australia

British Library Cataloguing in Publication Data

Windmill, Veronica
Human growth and development.
1. Human growth
I. Title
612.6 QP84

ISBN 0–7131–7490–0

Text set in 10/11 pt English Times
by Colset Private Limited, Singapore.
Printed and bound in Great Britain
by The Bath Press, Avon.

Contents

Preface

In writing this book, I have aimed to provide a readable coverage of human growth and development for those fairly new to the subject. I hope it will broaden knowledge of, and interest in, basic health and body care – with a long-term view to increasing awareness of people's needs, both as individuals and as members of society.

The book is in sections covering the main issues of human behaviour and development from childhood and adolescence through to adulthood and old age. At the end of each section is a series of assignments which encourage the students to discuss various issues, to make full use of audio-visual equipment, to use information-retrieval skills and observation in their own research work, and to study many different topic areas. The assignments can be used to reinforce course work covered during teaching time, to assess the student's work, or to ensure that the student has understood the topic.

Suggestions on where to find outside speakers, places to visit, and names and addresses of organisations that offer relevant advice and information are also included.

Overall, it is hoped that the approach will encourage students to take responsibility for their own learning and will allow the teacher to plan a more student-centred scheme of work.

Veronica Windmill

Acknowledgements

The publishers would like to thank the College of Health for permission to reproduce fig. 1.3, based on the 'Anatomy quiz' in *Self Health*, March 1984, and the Controller of Her Majesty's Stationery Office for permission to reproduce the ante natal record card (fig. 3.2), which is Crown copyright.

1 The body

In this chapter we discuss the functions of the major parts of the body and look at where they are situated.

It is important that we understand how our bodies work so that we can keep ourselves as healthy as possible and recognise when something is wrong. When we do become unwell, it is better for us – and more helpful to the doctor – if we can explain clearly what is wrong.

As well as helping us in our everyday lives, an understanding of the human body will also help in understanding the other sections of this book.

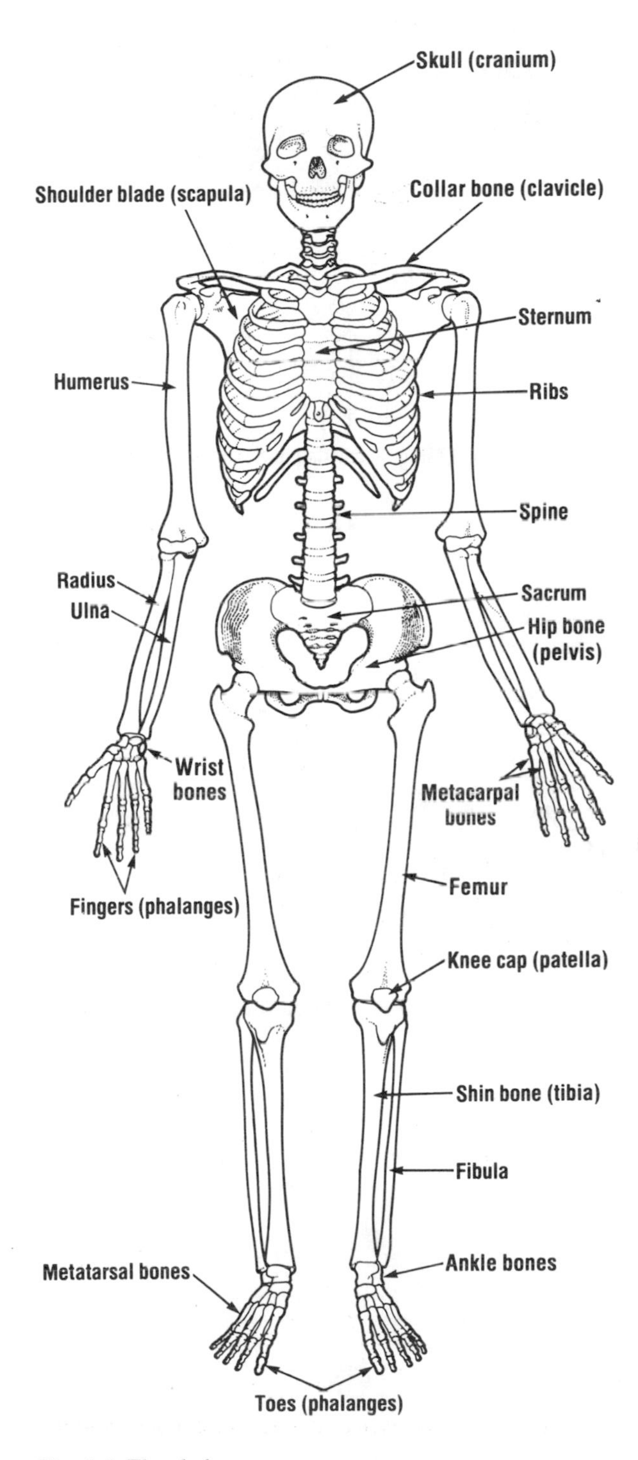

Fig. 1.1 The skeleton

Bones and the skeleton

The basis of the body is the skeleton (fig. 1.1), which is made up of strong tissue called bone. The centre of each bone is soft, as it contains bone marrow, and the bone becomes harder towards the outside. Like all parts of the body, bones grow. The speed of their growth can be affected by hereditary factors (which are inherited from the parents), hormones, and diet.

The skeleton has three main functions:

1. *Support* Without bones our bodies would be a floppy mass of tissue, unable to move. Our vital organs would crush each other, and we would die. The main support in the body is the spine, which ultimately supports our entire body.
2. *Protection* Our delicate vital organs are encased by a protective layer of bone. The brain is surrounded by the skull, and the heart and lungs by the ribs.
3. *Movement* Where two bones meet in the body, there is a joint which allows varying degrees of movement and flexibility.

The 206 bones in the human skeleton are held together by *ligaments,* which are tough bands of tissue that prevent the bones from moving too far.

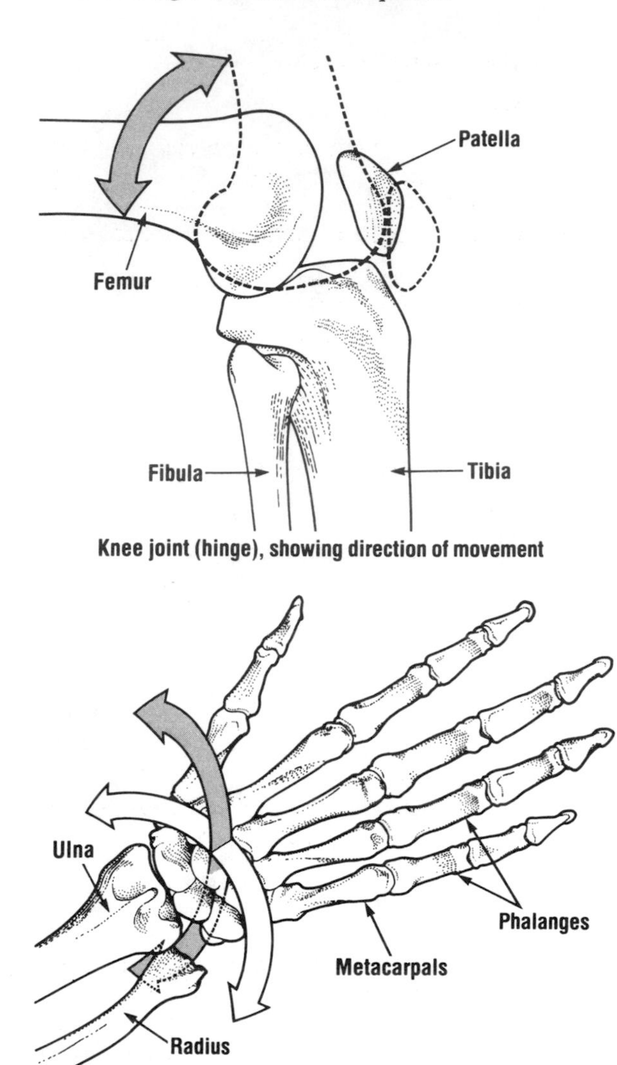

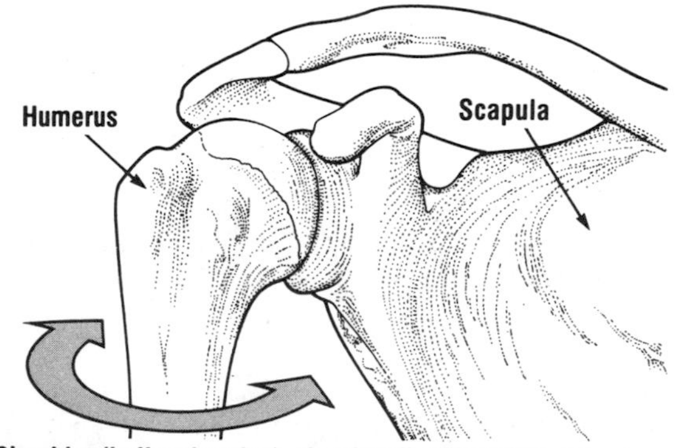

Fig. 1.2 The three types of joint

Joints and muscles

The places where our bones meet are called *joints*. Joints allow a wide variety of movements (fig. 1.2).

A *hinge joint* allows movement in one direction only, rather like a door hinge. This type of joint is found in the elbow, knee, and ankle.

The *double hinge joint*, such as the wrist, allows movement in two directions.

The most complex type of joint is the *ball-and-socket joint*, found in the hip and shoulder, where there is movement in many directions.

In order for our bones and joints to move, we need muscles. These are made up of tissue that contracts when stimulated by nerve impulses from the brain. Muscles need plenty of oxygen in order to work properly. If there is not enough oxygen getting through to the muscles, such as in a fast sprint, the muscles will get muscle fatigue, which is very painful. In some aerobic exercises, we are encouraged to carry on exercising until we feel a pain – the 'burn' – but there is some doubt as to whether this is good for the body.

The ends of the muscles are attached to the bones either directly or by bands of tough tissue called *tendons*.

In the finger joints, the tendons are long, because there is not enough room for bulky muscles on the fingers themselves. The muscles that operate the fingers are found in the arm, and are connected to the fingers by tendons.

The way that muscles work is quite complicated. They operate in pairs, as each muscle can only contract and relax, but not lengthen. Each muscle in the pair works opposite its partner so that, as one muscle contracts, the other relaxes and vice versa. Even when the body is at rest, the muscle pairs still operate in order to keep the body in position.

So far we have discussed the muscles found on the skeleton itself – the *skeletal muscles*. These are also known as *voluntary* muscles, as we have control over them and can decide when to move them. However, there are other muscles inside our bodies over which we have no conscious control, and these are called *involuntary muscles*. These muscles are found in many of our internal organs, such as the digestive system, the uterus, the bladder, and blood vessels. One of the strongest of these muscles is the heart, which pumps blood around the body throughout our lives.

Co-ordination

Both the voluntary and involuntary muscles receive messages from the central nervous system, which is made up of the brain and the spinal cord. The messages are passed along the nerve fibres in the form of electrical impulses. The end result is the movement of the muscle.

It is important to remember that most of our muscle movements are inside our bodies, where a complex network of nerves sends messages from the central nervous system to the muscles in order for them to function.

Assignments on the skeleton

1. As a class, try to obtain a model of a human skeleton so you can look at and name the various bones and joints.
2. There are three types of joint in the body: (i) hinge, (ii) double hinge, (iii) ball-and-socket. Draw a simple diagram of how each works and decide which of the three categories our joints fall into.
3. Look up these terms: endoskeleton, exoskeleton, vertebrate, invertebrate. Find out what each of these terms means, and then give three examples of each from the animal world.
4. Explain briefly the function of each of these bones: the skull, the ribs and sternum, the pelvic and pectoral girdles.
5. Choose one of the following and find out about it in more detail:

   a) The spinal column:

   - Draw a diagram and label it clearly.
   - Explain the make-up of the spine, its function, and how it affects the rest of the body.
   - Are there any joints in the spinal column? If so, what type of joints are they?
   - What is inside the spine itself?

   b) The limbs (both arms and legs):

   - Draw a diagram of each of the limbs and label them clearly.
   - What sort of joints are found in the arms and legs? Describe their uses.
   - Why are there so many tiny bones in the hands and feet?

Assignments on the muscles

1. In a group, discuss the following points:

   a) Why are some muscles under voluntary control while others are under involuntary control?

   b) Work out the co-ordination needed for some of these everyday activities:

   - cleaning your teeth
   - drinking a cup of tea
   - crossing the road
   - climbing the stairs
   - buttering a slice of bread

2. Individually, research the following terms:

   a) Reflex action:

   - What does this mean?
   - Name some of our body reflexes.

   b) Cartilage:

   - What is it?
   - Where would you find it in the body?

   c) Ligaments:

   - What are they?
   - Where are they to be found in the body?
   - What are their functions?

The organs

In the body, the term *organ* is used to describe a structure that has a special job to do.

Here is a list of some of the major organs in our bodies: heart, lungs, bladder, kidneys, stomach, eyes, ears.

Organs are grouped together to perform certain functions in the body. For example, between eating a meal and excreting the waste products of that meal some hours later, many organs have been used.

In the following section we shall be looking at the everyday functions carried out by our bodies. Before doing this, it is best to check that we can locate some of the major organs in our bodies by looking at the quiz (fig. 1.3).

Blood

Blood travels to all parts of our bodies. Without it we would die. Many people die from loss of blood. The British Red Cross Society says that if more

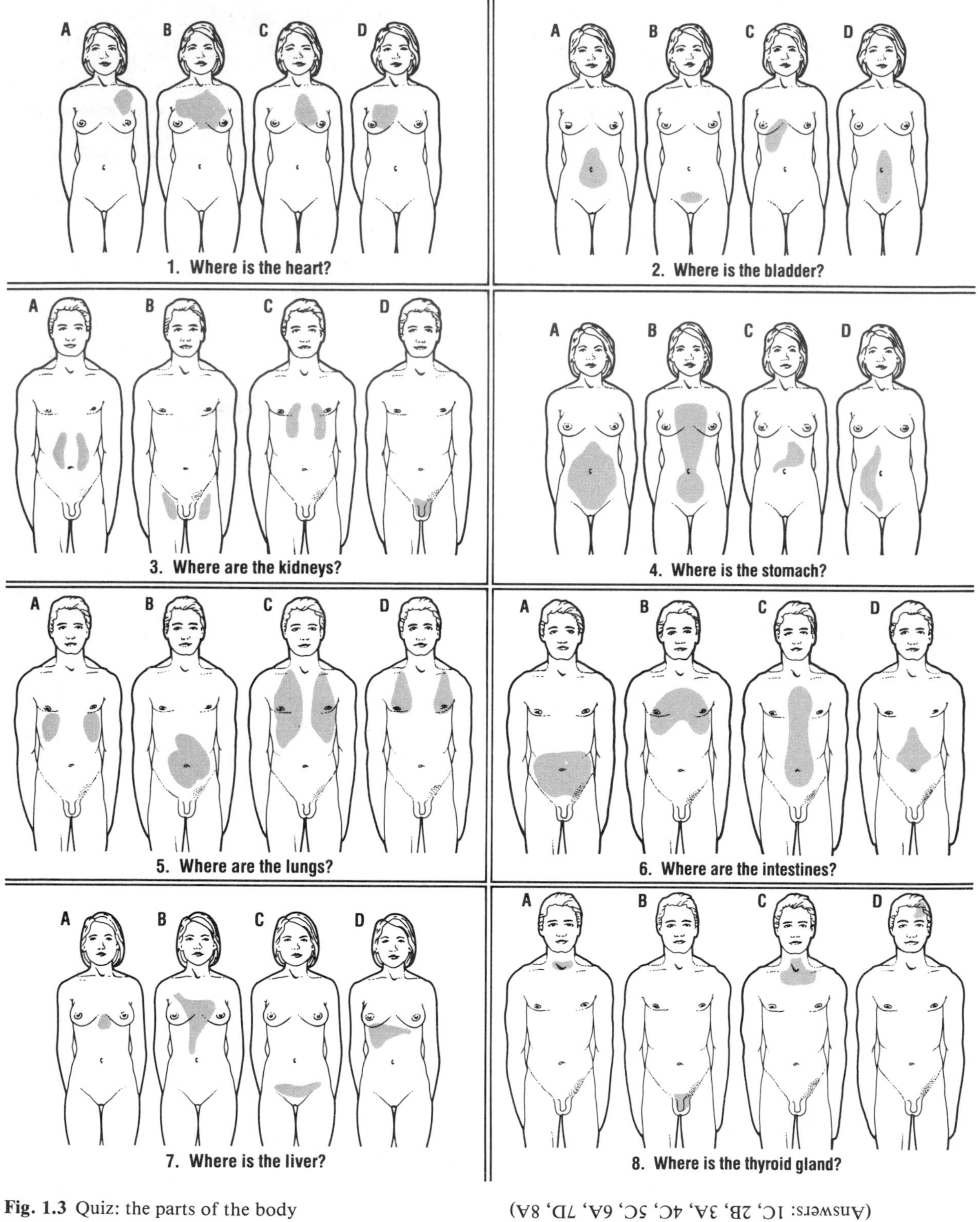

Fig. 1.3 Quiz: the parts of the body

(Answers: 1C, 2B, 3A, 4C, 5C, 6A, 7D, 8A)

members of the public knew how to stop bleeding at the scene of an accident, many lives would be saved.

Blood is carried around the body by means of the *circulatory system*.

The blood itself has seven main functions:

1. When we breathe air, we are taking oxygen into the bloodstream. This is called *respiration*.
2. When we eat, the digestive system breaks the food down into glucose and other substances, which are then transported by the bloodstream to feed the body.
3. Waste products from the body are carried away in the blood to be disposed of.
4. We need fluid in our body cells, and the blood carries this fluid around to where it is needed.
5. Our bodies are protected from infection by the action of the white cells in the blood. There are five different types of white blood cell, and three of these are directly responsible for fighting infection.
6. The blood carries the hormones we need to regulate our bodies (these are discussed later, on pages 13–14).
7. The blood helps to keep our body temperature reasonably steady (unless we are ill), so that in summer our bodies are the same temperature as in winter.

The blood is pushed around the body by the heart, which is a strong muscular pump. The blood from the heart is channelled into a series of tubes called *arteries*. These then branch out to supply the limbs and organs with blood. Once the arteries reach the limbs and organs they branch out into finer, smaller tubes called *capillaries*.

Once the body tissues have been given food and oxygen from the blood, the capillaries rejoin into larger tubes called *veins*, which take the blood containing any waste products back to the heart.

Respiration

Why do we need to breathe? The air we take in contains oxygen, which our bodies need, along with glucose from the food we eat, to make into energy.

Breathing, or respiration (fig. 1.4), is the method by which oxygen from the air is passed around the body to all the tissues and organs. The air we breathe out has more waste products in it than the air we breathe in. It has less oxygen, but more carbon dioxide and water.

The lungs, although they are very elastic, have no muscles of their own enabling us to breathe in or out. It is the muscles around the ribs which do this. Simply speaking, in order to breathe air in (inhale), the rib muscles lift and the diaphragm muscle

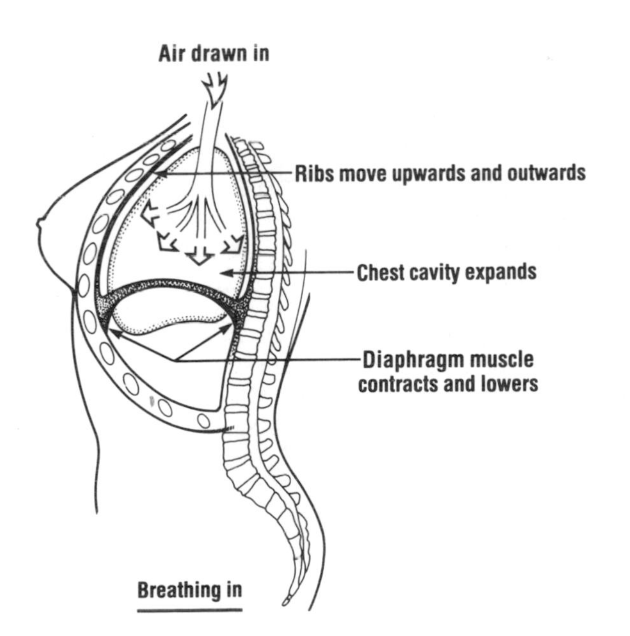

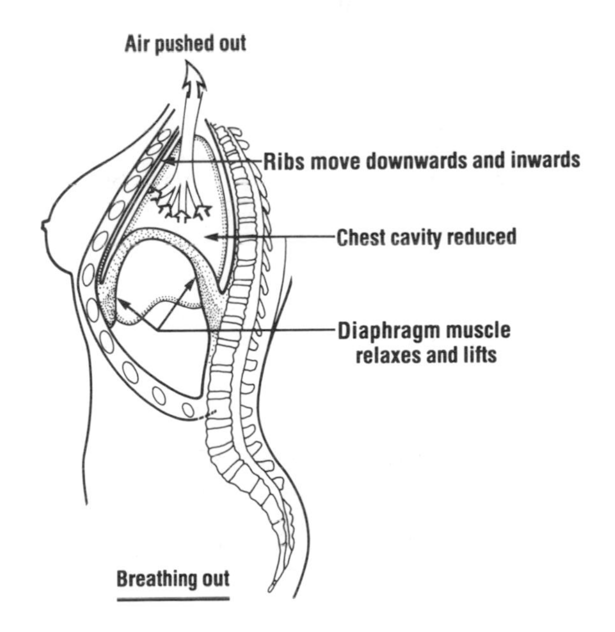

Fig. 1.4 Respiration

lowers to enlarge the space in the chest cavity while air is brought in through the nose and mouth. To breathe air out (exhale), the rib muscles lower and the diaphragm muscle lifts to push the air out.

What goes on inside the lungs is far more complex, as it is here that the oxygen from the air has to pass into the bloodstream while the carbon dioxide and water from the bloodstream are passed back into the air.

Assignments on the blood and circulation

1. Blood is circulated throughout the body in a series of tubes called *arteries* and *veins*. As a group, carry out the following tasks:

   a) Find a diagram or model which shows the major arteries. You will find these arteries named: carotid, aorta, genital, renal, radial, ulnar, femoral, tibial, iliac, brachial. The names suggest which part of the body the arteries are linked with. Discuss these links.

   b) The heart muscle pumps the oxygenated blood around the body through the arteries. The deoxygenated blood then returns along the veins. The blood leaving the heart is carried under great pressure, so the artery walls need to be thick and strong. Discuss the following points:

   - why pressure is needed to circulate the blood
   - why there is less pressure in the veins
   - why a cut artery is more dangerous than a cut vein
   - how blood flows in one direction only
   - why the blood in the arteries is bright red whereas the blood in the veins has a 'bluey' tinge

   c) Doctors can measure blood pressure using an instrument called a *sphygmomanometer*. Find out how the gauge works. Perhaps someone from your local health centre would be willing to give a demonstration to the group.

   d) Although we cannot actually see our hearts beating, it is possible to detect the heart rate by feeling for a *pulse*. There is a pulse every time the heart beats. The pulse is lowest when we are asleep and highest when we are exercising or excited.

   Work in pairs to measure pulse-rates. You need paper, a pencil, and a stop-watch. Before starting, make sure you can find a pulse, either at the wrist or just above the collar bone. Use the little finger to feel for a pulse, as the other fingers may be relaying your own pulse-rate. Take the pulse-rate for one minute for the following conditions:

   - at rest
   - after walking around the room for one minute
   - after ten jumping jacks
   - after jogging on the spot for one minute
   - after resting for two minutes

After collecting the data in pairs, compare pulse-rates between the members of the group, and discuss these questions:

- Why is it healthier to have a lower pulse-rate?
- Why does our pulse-rate rise when we exercise?
- Why does regular exercise lower the pulse-rate?
- Babies have a pulse-rate of 100–120 a minute, children of six to ten years of 60–80. Can you explain why this is?
- After exercise, a healthy heart returns to normal quite quickly. How long did it take yours to reach a normal pulse-rate?

2. Research into one of the following and report your findings back to the group in either a written or an oral report.

   - plasma
   - red blood cells
   - haemoglobin
   - white blood cells
   - platelets and blood clotting

   Remember to include in your report a definition of the item, a description of its composition, and its functions.

3. As with all parts of the body, sometimes things can go wrong with the circulatory system. Some of these problems are more serious than others. Look into the conditions listed below and find out the symptoms, the causes, and the cure (if any). As with question 2, report your findings back to the group.

   - anaemia
   - angina pectoris
   - cardiac arrest
   - coronary thrombosis
   - fainting
   - haemophilia

- high blood pressure – hypertension
- leukaemia
- palpitations
- varicose veins
- bone-marrow disorders in children

4. Recently, there has been a great deal of publicity about keeping our hearts and arteries healthy by exercising regularly and eating a well-balanced diet. Contact your local Health Education Council office and arrange to collect some of the many leaflets it publishes about this. Read the leaflets and check whether you and your family are doing the right things.
5. Until as recently as the mid-nineteenth century, blood-letting was carried out for so-called medical reasons. This practice involved cutting open the veins and removing up to a litre of blood. It was thought that blood-letting would calm the nerves, take away bad temper, cure illnesses and keep the body healthy. Sometimes, human blood was thought to increase the fertility of the land. After blood-letting came the practice of using leeches to suck out the blood.

   If you wish, find out more about these practices and discover in what way they are alien to today's medical thinking.
6. As a contrast to the old idea of blood-letting, today we carry out blood transfusions which save lives by giving blood to people who have lost more than their bodies can replace. Find out when the Blood Transfusion Service is next visiting your area (its telephone number is in the directory). Perhaps a representative would be willing to come and discuss its role with the group.

Assignments on respiration

1. How does the oxygen from the air reach the bloodstream? As a group, discuss how this complex process works.
2. The air we breathe out contains more water and carbon dioxide than the air we breathe in. Think of everyday happenings that prove this point.
3. The lungs need to be kept clean in order to function efficiently. The lungs are protected by the larynx and epiglottis, coughing, cilia, and mucus. In pairs, find out about each of these. Include the following points:

   - a definition of each
   - where it is found
   - how it protects the lungs

4. As a group, find out this information about the lungs:

   - Where are the lungs found in the body?
   - Approximately how much do they weigh in an adult?
   - How often do we breathe at rest?
   - Why do we breathe more frequently during exercise?
   - How are speech, laughter, yawning, hiccups, sneezing, and crying caused?

5. As a group, discuss the dangers in these conditions:

   - high altitude, where there is less oxygen than at sea level
   - below sea-level, where there is more pressure
   - a heavy weight pressing on the chest, such as being buried in an avalanche, or even under sand on the beach
   - choking

6. In pairs, research into some of these respiratory problems and diseases. Report your findings to the group in a written or an oral report. Make sure all are covered by the group.

   - asthma
   - acute and chronic bronchitis
   - colds
   - hay fever
   - influenza
   - laryngitis
   - lung cancer
   - pleurisy
   - pneumonia
   - tuberculosis

   In each case cover the symptoms, the causes (if known), and any remedies and cures that are available.

Digestion and nutrition

The food we eat needs to be converted into a form that can be transported around the body in the bloodstream. This process is called *digestion* (fig. 1.5).

The process of digestion is better explained as a series of steps.

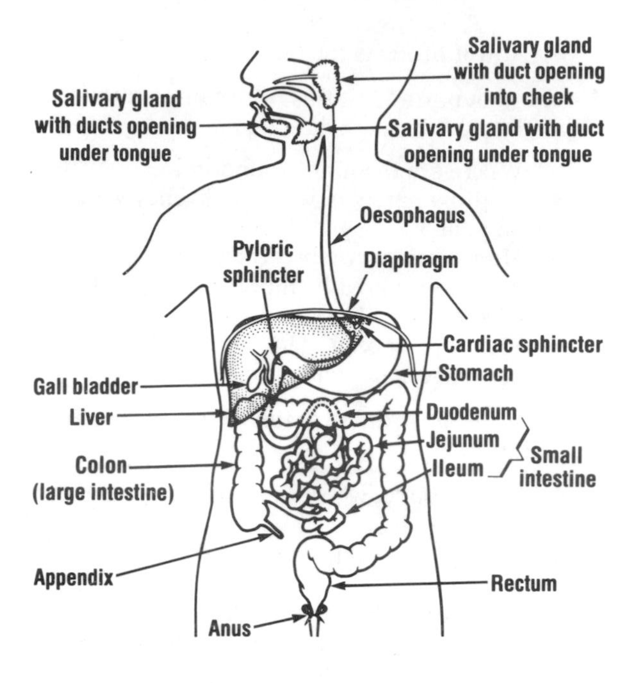

Fig. 1.5 The digestive system

1. The process starts in the mouth, where the food is chewed and mixed with saliva.
2. It is swallowed into the stomach, which is a strong muscular bag with an opening at the top and bottom. These two openings are kept closed by strong muscular rings called *sphincters* – the *cardiac sphincter* at the top and the *pyloric sphincter* at the base. After swallowing, we have no further control over what happens to the food we have eaten.

   During the two to three hours the food is stored in the stomach, it is pounded and churned by the muscular walls until it becomes semi-fluid. Meanwhile, the gastric juices in the stomach break down the food even more.
3. The pyloric sphincter gradually lets out small amounts of the mixture into the *small intestine*. The small intestine is coiled up inside us as it can be as long as 7 m in an adult. It is divided into three parts:

   i) the duodenum, where the digestive juices from the liver, pancreas, and intestinal walls help to digest the food
   ii) the jejunum
   iii) the ileum, where much of the digested food is absorbed
4. After much of the partially digested food is absorbed, what is left passes on into the large intestine. The contents are mainly water, roughage, bacteria, and mucus. Water is absorbed through the walls of the large intestine and what is left becomes the faeces: the waste product of the food we eat. These are passed out of the body through the anus.

Our bodies need food to give us energy to survive, so what we eat is very important, and may affect the efficiency of our body's performance.

Nowadays, there is a great deal of emphasis on healthy eating, not only in order to keep slim, but also to prevent heart disease, high blood pressure, and so on.

In addition to water, we need five basic nutrients:

1. carbohydrates for energy
2. proteins for building muscles and other tissues
3. fats for energy and heat insulation
4. vitamins to help various body processes
5. minerals to help form bones and blood, and to aid energy transfer
6. water to help with chemical reactions and waste disposal

1. *Carbohydrates* These are mainly sugars and starches which are found in sugar, cereals, fruit, vegetables and nuts. Ideally, our carbohydrate intake should be from fresh fruit and vegetables and unrefined cereals, rather than from sugar and refined cereals.
2. *Proteins* Proteins should be no more than a quarter of our total food intake, and preferably less. They are found in meat, fish, and dairy products as well as in pulses (such as peas, lentils, etc.).
3. *Fats* These should make up less than a tenth of our daily food intake. Research has shown that vegetable fats (such as nut oils and some cereal oils) are better for us than animal fats made from meat and dairy products.
4. *Vitamins* All our food except sweets, crisps, white sugar, jams, ice-cream, squashes and other highly refined foods contain various vitamins. The vitamins essential for good health are A, B, C and D.
5. *Minerals* Minerals such as calcium, phosphorus, iron, iodine, sodium, potassium, magnesium, and so on are all needed in minute quantities and are naturally present in a good diet.

Assignments on digestion

1. As a group, check that you understand what are meant by the following terms: saliva and other digestive juices, stomach and sphincters, small intestine (the duodenum, jejunum and ileum), large intestine, alimentary canal, peristalsis.
2. Digestive juices are necessary to break down the food we eat. In more detail, find out about the digestive juices produced by the following: mouth, stomach, liver, pancreas, small intestine.
3. In pairs, research into these problems and diseases associated with the digestive system. Report your findings to the group orally or in a written form.

   - appendicitis
   - colostomy and ileostomy
   - constipation
   - diarrhoea
   - dyspepsia (indigestion)
   - flatulence
   - gall-stones
   - gastroenteritis
   - haemorrhoids (piles)
   - hiatus hernia
   - ulcers (peptic: duodenal and gastric)

   In each case, try to explain the symptoms, causes, and cures.
4. As a group, check that you understand the meaning of these terms: obesity, carbohydrates, proteins, fats, minerals, vitamins, refined and unrefined foods, 'whole' foods, roughage and fibre, calorie (as a measure of energy), metabolism.
5. Contact your local Health Education Council and arrange to collect the leaflets that are available on healthy eating. After reading these leaflets, discuss the following as a group:

   - the benefits of unrefined foods
   - the need for fibre in our diet
   - the dangers of refined foods and sugar
   - why we should eat less salt
   - why we should eat less fat
   - the hidden dangers in 'junk foods'

6. Eating more calories than our bodies need leads to obesity. In pairs, find the number of calories needed by the following: a manual worker, a sedentary worker, a man, a woman.

   Obtain a list showing the calorie content of various foods, and work out how many calories you eat on a normal day. Without going on a strict diet, what would be the best way to prevent weight gain in our everyday eating?
7. Working individually, copy out this table showing the vitamins A_1, B_1, B_2, B_6, B_{12}, C, D, E, K. Fill in the columns.

Uses	*Food source*	*Signs of deficiency*
A_1 B_1 etc.		

8. The rate of our metabolism can affect the speed at which we 'burn up' the food we eat. Someone with a low metabolic rate needs less food than someone with a high metabolism.

   As a group, discuss your individual daily food intake and the amount of exercise you take. Are any members of the group able to 'eat anything' without putting on weight? Do some members eat very little and still feel they are overweight? Are the other members of your families of a similar build?
9. Food additives are beginning to cause concern, as research has shown that many people may be allergic to them. As a group, look at the list of ingredients in a variety of food items, and see what additives are included. Find out which additives are colourants, preservatives, emulsifiers, and stabilisers, and which are likely to cause health problems.

   Useful sources of information: *E For Additives* by Maurice Hanssen, published by Thorsons; the Health Education Council; dieticians; the local library.
10. Convenience foods can be useful and time-saving. Yet many are also costly and high in additives, fats, sugar, and salt. As a group, discuss your views on convenience foods, including some of these points if you wish: cost, time, taste, appearance, nutritional value.
11. Even if we prefer home-cooking, it is often necessary to eat out sometimes, such as when we are at work or school, on holiday, in hospital or a residential home, receiving meals-on-wheels, and so on. Over a week or so, look at some of the menus provided at a canteen. This

needs to be carried out tactfully in order to avoid the management or staff feeling they are being criticised. Analyse the menu, asking these questions:

- How much of the food was fried?
- Was there a salad alternative?
- Was there a choice?
- Was the diet nutritionally balanced?
- Was there plenty of fibre?
- Were any convenience foods used?
- What prices were charged?
- Was there any guidance about healthy eating?
- Was fresh fruit available as an alternative sweet?

12. It is possible to eat a healthier diet without having to change our eating habits drastically. As a group, discuss alternatives to or healthier ways of preparing these foods:
    - fried streaky bacon and fried egg on fried white bread
    - cornflakes and white sugar with silver-top milk
    - battered fried fish and chips
    - tinned fruit in syrup with cream
    - white bread and butter
    - crisps or biscuits as a snack
    - baked potato with butter topped with cheddar cheese

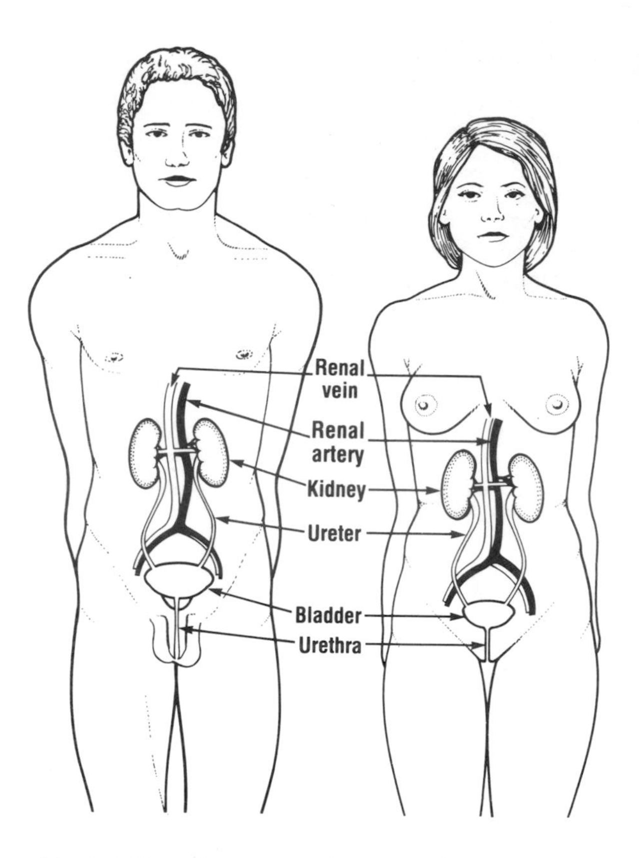

Fig. 1.6 The urinary system in men and women

Excretion

All the body processes result in some kind of waste product, which has to be disposed of in order to keep the body running smoothly and efficiently.

The waste product of eating passes out of the body in the form of faeces. The waste products of respiration are breathed out in the forms of carbon dioxide and water.

There are also waste products from the internal body processes, and these also need to be removed from the body. This is mainly carried out by the kidneys, but the skin and liver also have a role to play.

The kidneys produce urine as a waste product, and this is collected in the bladder. The urinary system (fig. 1.6) is closely linked to the reproductive system, even though they have entirely separate functions. Doctors specialising in urinary problems also tend to specialise in reproductive problems.

We have two kidneys, which act as filters. Here the waste products are removed from the bloodstream, and any essential substances are reabsorbed. These waste products become urine, which is then passed along a tube called the ureter and collected in the bladder (fig. 1.7). The average person passes about 1 – 1 ½ litres of urine a day.

The bladder is a muscular bag capable of stretching enough to hold the urine, which is usually a little over ¼ litre.

When the bladder needs emptying, waves of contraction are felt by the individual. Under normal circumstances, the person will wait until it is convenient before emptying the bladder. At the right time, the bladder relaxes its sphincter and the urine is released by muscle contraction.

Excretion through the skin is carried out by the sweat glands, which expel some waste products in the form of sweat. Ninety-eight per cent of sweat is water. Sweat glands are found in the groins, arm

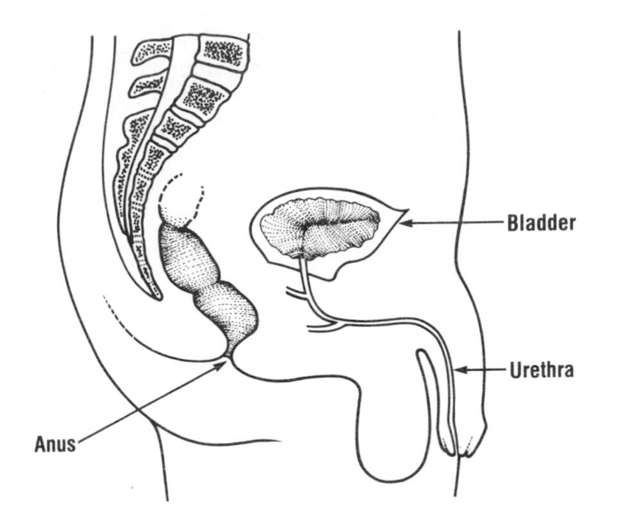

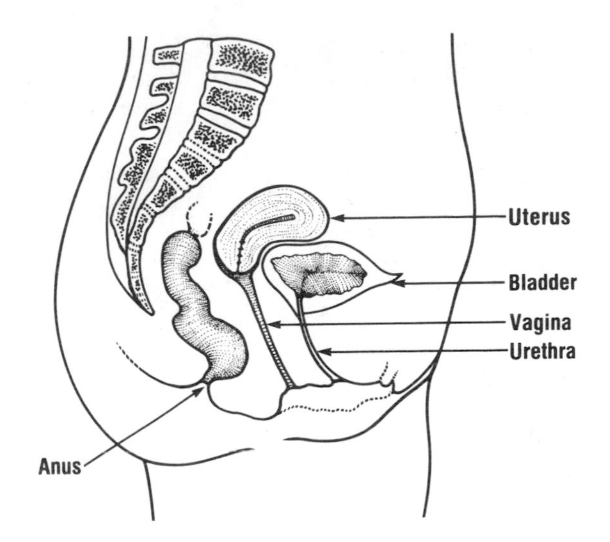

Fig. 1.7 Position of the bladder in men and women

pits, forehead, palms of the hands, and soles of the feet.

As well as removing waste products, sweat is essential for keeping our body temperature even. If we become too hot, either through exertion or fever, we produce sweat which then evaporates from the skin surface and reduces the body temperature.

Assignments on excretion

1. As a group, try to obtain an enlarged model of a kidney to see how the blood is filtered and the waste products collected. Alternatively, cut an ox kidney in half longways.
2. Try to explain these observations:
   - In the morning, urine is dark coloured and has a stronger odour.
   - After drinking alcohol, or large amounts of tea and coffee, we need to pass urine more frequently.
   - Some children are 'potty-trained' earlier than others.
3. Bed-wetting, or enuresis, may occur at any age. What are some of the causes of bed-wetting? How can the problem be reduced or prevented?
4. Cystitis is an inflammation of the bladder. The Health Education Council produces some helpful leaflets on the subject. With the help of these leaflets, find out about cystitis.

The sense organs

We have five main senses: sight, hearing, smell, taste, and touch. With our senses we learn about what is around us, and can protect ourselves from any dangers. Obviously, there are more senses than these – there are sensitivity to pain, temperature awareness, and balance, as well as many others – but in this section we are going to discuss the eye, the ear, taste, and smell.

The eye

The eye (fig. 1.8), as well as being the organ of sight, is also part of the brain – the only part that is not protected by bone. As well as sending information to the brain from the outside world, the eye can give information about what is happening inside the brain.

The eyeball itself is fairly small, weighing about 7 grams and measuring about 2.5 cm in diameter. The *iris* is the coloured part of the eye and, at birth, light-skinned babies have blue eyes, as there is little colour pigment present. As the baby grows, the colour of the eyes will change to green or brown in many cases. Dark-skinned babies have greyish-green eyes at birth, changing to brown later. After death, the pigmentation again changes and all eyes become a greeny-brown in colour.

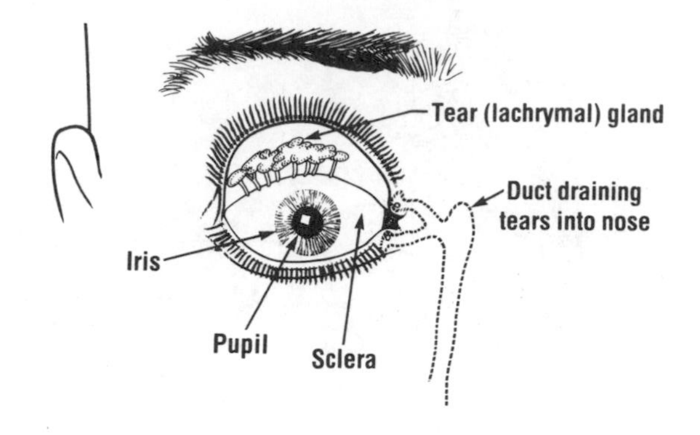

Fig. 1.8 The eye, with eyelids pulled back to show position of glands and ducts

The circular opening inside the iris is called the *pupil*, and it is able to dilate or contract. If the light is bright, the pupil will contract to keep out some of the brightness. If it is dark, then the pupil will dilate to allow as much light in as possible.

We are able to move our eyes because there are muscles behind the eyeball. Usually these muscles pull together, unless there is some problem. In most cases any problem can be corrected by treatment or surgery.

The eye is quite well-protected by the eyelids and eyelashes, which prevent foreign bodies from getting into the eye. Tears, which are produced in the tear glands, keep the eye well-lubricated and help to prevent any infection.

The ear

The ear (fig. 1.9) picks up the sounds going on around us and transmits them to the brain. The part of the ear we can see is called the *outer ear*, which collects the sound and leads to the *ear drum*. The sounds will make the ear drum vibrate. Inside the head, and behind the ear drum, is the second part of the ear, called the *middle ear*, which contains three little bones (see diagram). These pick up the vibrations from the ear drum, amplify them and send them on to the third part of the ear, called the *inner ear*. Here, the auditory nerve picks up the sound messages and sends them on to the brain.

Taste and smell

These two senses are strongly linked, as anyone who has had a bad cold will know – if you can't breathe and smell, you can't taste.

We identify flavours through our *taste-buds*, which are mainly found on the tongue, though there are a few on the palate and in the throat. Each taste-bud can distinguish either sweet, sour, salt, or

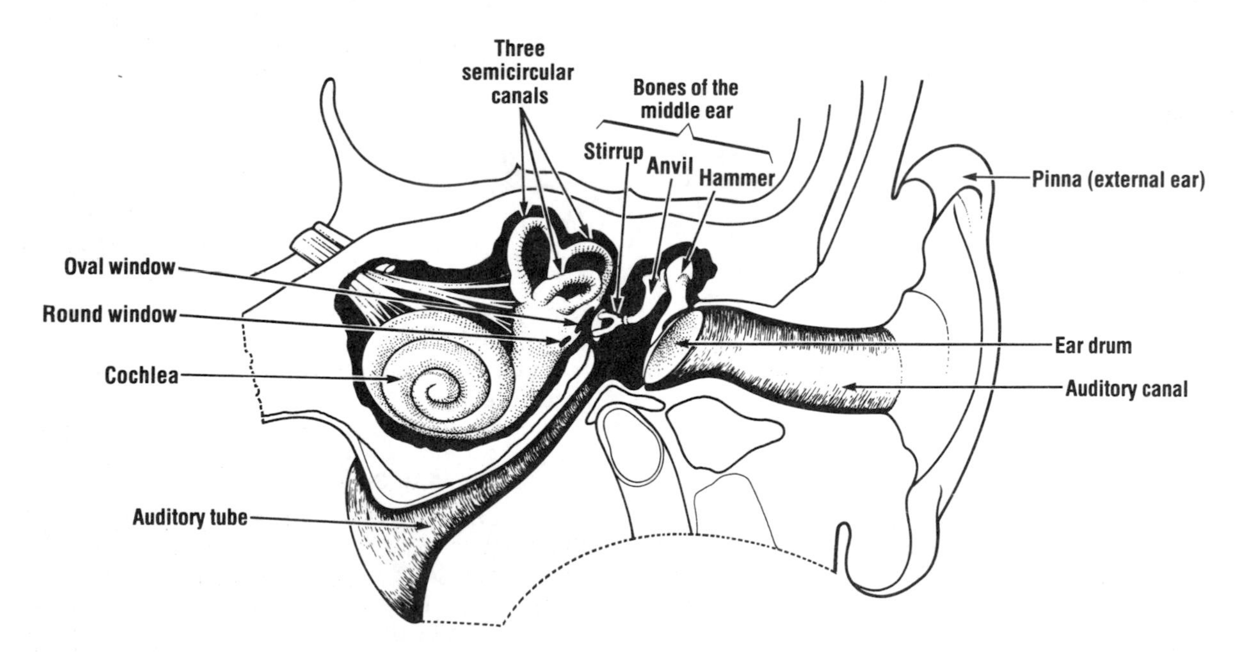

Fig. 1.9 The ear

bitter flavours, and certain areas of the tongue seem to be more receptive to each of these four taste categories.

Smell receptors lie in the nasal membrane just above the stream of air we breathe in. The sense of smell is very sensitive; it is thought we can distinguish about three thousand different odours.

Assignments on the senses

1. As a group, find out how the eye is able to see. It may help to obtain one of the models available to show how this works.
2. There are various disorders that can affect our eyesight. Find out the symptoms and causes of these disorders:
   - long sight (hypermetropia)
   - short sight (myopia)
   - astigmatism

   What can be done to correct these disorders? Ask a local optician if he or she would be willing to talk to the group about this and how eyesight is tested.
3. In pairs, investigate the causes of the following:
   - pupils dilating and contracting
   - styes
   - bloodshot eyes
   - squinting
   - colour blindness
4. As a group, contact the Royal National Institute for the Blind. It may be able to put you in touch with someone in your area who would be willing to talk to the group about the problems of being blind, and show some of the aids available for the blind.
5. Find out what there is in your area for the blind – include any voluntary or statutory agencies and services.
6. As a group, check that you understand how the ear works and can pick up the sounds around us. Again, a large model may be of help.
7. In hospitals there is an ear, nose and throat department. Draw a diagram that will show why these three areas are closely related.
8. As a group, find out this information:
   - Why do ears 'pop' when we change height rapidly, such as in an aeroplane? Why are we given sweets to suck to help clear our ears?
   - Why do we have wax in our ears?
9. Contact the Royal National Institute for the Deaf and ask whether it can suggest someone to talk to the group about the causes of deafness and the problems of being deaf.
10. Find out about the deaf sign language.
11. What is available in your area to help children and adults with hearing disabilities?
    - Is there a hearing-impaired unit attached to the local school? If so, see if there is someone to tell you how a child is helped to cope with this handicap.
    - How often are hearing tests carried out? (A health visitor may help you here.)
    - What help is available for the hearing-impaired adult?
12. Experiment with your taste-buds. Find various foods that are sweet, sour, salt, and bitter. Taste them with a blindfold on. Can you recognise the flavour? Can you tell immediately whether it is sweet, sour, salt, or bitter? Is there any one part of your tongue that reacted to each of the four tastes?

Nerves and hormones

Our bodies would be useless if the organs inside them had no way of communicating with each other. Messages from one part of the body frequently need to be sent to another part. For example:

- During vigorous exercise the muscles need more oxygen from the blood to keep going. To supply this, the heart needs to pump blood around the body faster, so the body responds by increasing the rate of breathing.
- If we lose our balance we immediately reach out for something to stop ourselves from falling over.

These messages are passed on by the *nervous system*, which is made up of the brain and spinal column, and many nerve fibres. Messages from the environment outside the body, or from inside the body, are passed on very quickly and efficiently by nerve impulses. The nervous system is under the control of the brain. Messages to and from the rest of the body are carried along the spinal cord.

Another method of sending messages inside the body is by hormones, produced by the *endocrine system*. Hormones take longer to transfer messages than the nervous system, as they are carried along in the bloodstream, but the effect may last longer.

Assignments on nerves and hormones

1. All the endocrine glands listed below produce hormones that affect the way our bodies function. Find out what hormone each gland produces and describe the effects of each hormone. Also, discuss the possible effects of a malfunction in any of these hormones:
   - pancreas
   - adrenal glands
   - thyroid gland
   - parathyroid gland
   - pituitary gland
2. What reflexes are there in the human body? Working in pairs, try to test some of your partner's reflexes.
3. Research some of the causes of brain damage and paralysis.
4. Find out the symptoms, causes and possible treatment of
   - migraine
   - concussion

2 Sex and pre-conceptual care

Understanding human sexual relationships

Humans are unusual in the animal world because they mate throughout the year, whereas other animals have a particular mating season, usually when the female is 'on heat'. After this mating the female will reject the male, yet in humans the female is receptive to the male at any time, even during pregnancy. This suggests that the human sexual act is more than merely a means of reproduction. This belief is further confirmed by the fact that women are able to have orgasms; there is no female equivalent in the animal world. One theory to explain this feature is that humans usually mate facing each other, which may stimulate the woman to orgasm, whereas animals mate from the rear. However, this does not explain the fact that humans make love for pleasure and intimacy as well as to make babies. In the animal world, the sex act may only take a few seconds to complete. Although this may be true for some human couples, we tend to take time over it. In addition to this is the frequency of love-making: there is no doubt that the average couple make love more frequently than animals.

With human sexual relationships, it is generally agreed that love-making is a personal thing and, as long as both partners are happy with their sex lives, what they do and how frequently they do it is up to them. However, there are many general observations that can be made about human love-making. First of all, it is important to understand the external and internal appearance of the male and female reproductive organs.

The female reproductive organs

The female reproductive system is both inside and outside the body. The sex organs outside the body are called the *vulva* (fig. 2.1), and are made up of the inner and outer *labia*, which are two folds of flesh. The inner labia join together at the front and enclose the *clitoris*. The clitoris is made of spongy,

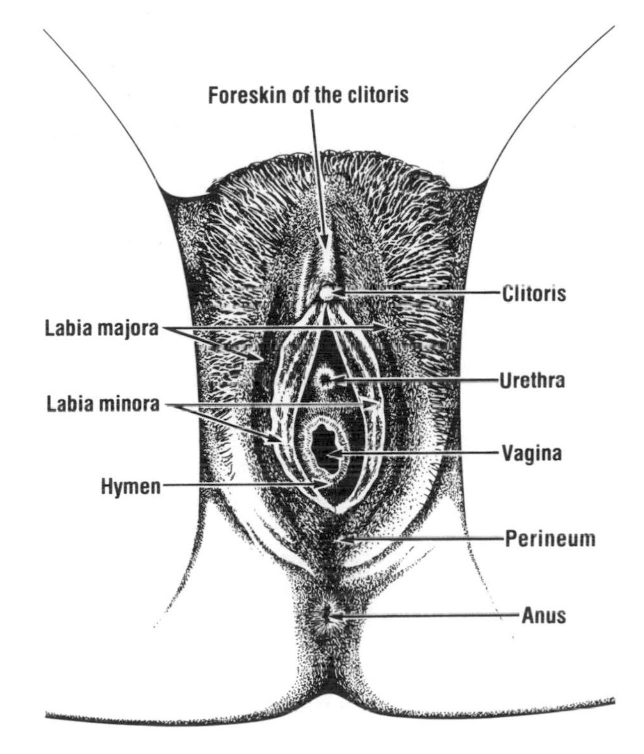

Fig. 2.1 The vulva

erectile tissue rather like the male penis. Just behind the clitoris is the opening to the bladder, the *urethra*, and behind that the *vagina*. The vaginal passage is 10–12 cm long, and is the start of the internal sex organs. It has ridged, elastic walls to allow the vagina to stretch during childbirth or sexual intercourse. The opening to the vagina is covered at birth by a fold of skin called the *hymen*. As the girl grows older, the hymen may be broken naturally by exercise, or the use of tampons.

The pear-shaped *uterus*, or *womb*, is at the top of the vagina (fig. 2.2). The neck of the womb is called

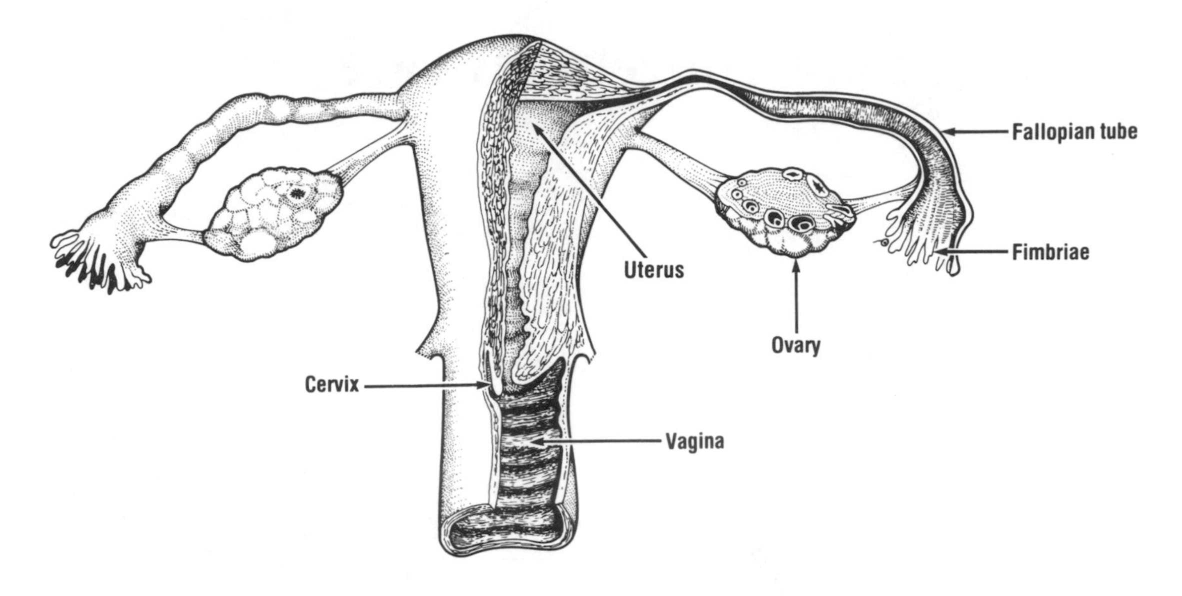

Fig. 2.2 Female reproductive organs

the *cervix*. At the wider top part of the womb, two *Fallopian tubes* lead off to the ovaries. The ovaries are walnut-sized and produce the female hormones oestrogen and progesterone, as well as the eggs (*ova*). A baby girl is born with all the eggs already formed in her ovaries (around 200 000 ova).

At puberty, the ovaries start to develop ripened eggs, usually one a month. When the egg is completely ripe, it bursts out of the ovary and enters one of the Fallopian tubes. This process is called *ovulation*.

While the egg journeys on to the uterus, the lining of the uterus thickens to prepare itself for a fertilised egg to be implanted; the blood supply is increased in readiness. If no fertilised egg is implanted in the lining of the uterus, the lining together with some blood and the unfertilised egg pass out of the uterus through the vagina about 14 days after ovulation. This process is called *menstruation* or, more commonly, a *period*. Once the period has finished, after about 5 days, the entire cycle starts again. Although 28 days is considered to be the normal length of the cycle, this varies considerably from one woman to another. There is only cause for concern when a woman's individual cycle alters.

The male reproductive organs

The man's sex organs are on the outside of the body and are made up of the penis and testicles (fig. 2.3). The penis, normally soft and spongy, has the capacity to become erect in the same way as the female clitoris. The top part of the penis, called the *glans*, is covered by the *foreskin*. Sometimes a boy will have an operation called *circumcision*, which removes the foreskin. Most circumcisions are carried out for religious or cosmetic reasons, but sometimes it might be necessary for medical reasons, such as a tight foreskin. The foreskin is intended to protect the delicate glans.

The shaft of the penis leads down from the glans to the testicles, which are in a bag of skin called the *scrotum*. The scrotal sac, which is sparsely covered in hair, is able to shrink when cold or sexually excited. Inside it is divided into two, each half containing a testicle (or *testis*). Each testicle is about the size of a walnut – rather like the female ovaries – and in most cases the left hangs slightly lower than the right. The testes produce hormones and sperm. They hang outside the body because sperm production is more successful at a temperature slightly lower than body heat.

The sperm are made in a series of fine tubes

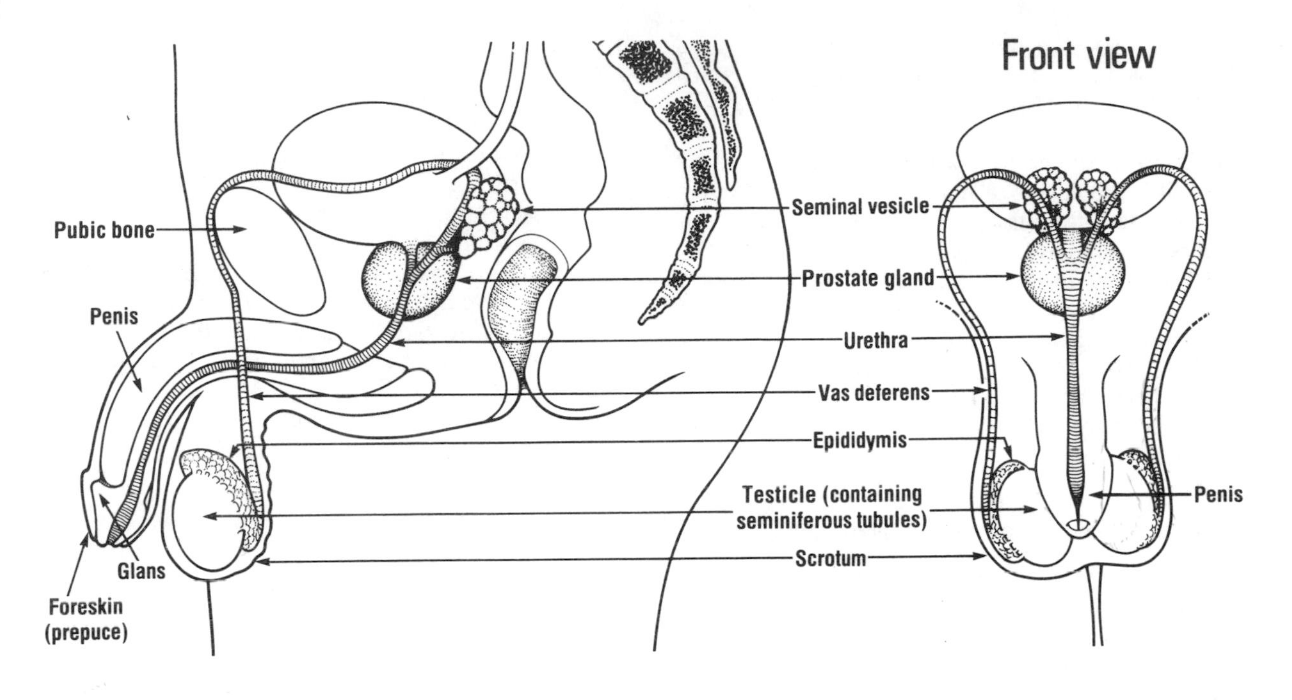

Fig. 2.3 Male reproductive organs

called *seminiferous tubules*, and are then stored in the *epididymis* until needed. Before being ejaculated, the sperm travel from the epididymis along a tube called the *vas deferens* until they reach the *prostate gland*. Here they are mixed with seminal fluid and become *semen*, which then passes along the urethra and penis before being ejaculated. Each ejaculation contains at least 300 million sperm. If the sperm are not ejaculated within a few weeks, they are broken down by the body and reabsorbed.

The male penis, unlike the female vagina, has two functions: it is also the urine outlet. During sexual arousal, the normally flaccid penis becomes larger and erect because of an increase in the blood supply. The muscles at the base of the penis tighten up to prevent the blood from escaping, so the penis remains erect. The opening to the bladder closes to create a clear passage for the sperm.

Sexual intercourse usually sounds very cold and clinical when described in a book, but in most cases there will be love, or at least deep affection, between the partners. In a good sexual relationship, the orgasm is not the be-all and end-all of love-making. The couple will spend time on what is called *foreplay*, where they may kiss and cuddle and generally express their feelings for each other to make the act one of shared intimacy and mutual pleasure.

When she is sexually aroused, the woman's vagina becomes lubricated by mucus secreted by the vaginal walls, and her clitoris becomes erect. The man's penis will also become erect, and a small amount of lubricating fluid is produced.

After foreplay, the couple will be ready for penetration, which is when the penis enters the vagina. Once the penis is inside, the couple's sexual arousal will increase because of the frictional movements: their pulse-rates will increase and their blood pressure will rise until the point of orgasm is reached. The man's orgasm causes the ejaculation of his sperm, but for both the orgasm is very pleasurable in itself. In most cases, a man will have one orgasm, but women are able to have more than one.

It is up to the couple to find out what each of them likes during love-making. Experience, love and consideration play a part in ensuring that both partners are satisfied.

Intercourse seems on the surface to be quite a simple act, but it can be a source of great anguish to couples with sexual problems. Sometimes these problems can be helped if the couples learn to discuss them openly with each other, but they may need professional help to overcome their difficulties.

Assignments on human sexual relationships

1. Find out about the symptoms and treatment of these common sexual problems:

   - premature ejaculation
   - impotence
   - frigidity
   - loss of libido (interest in sex)

   Remember to include possible causes of the problem, and where the person could go for help.
2. As a group, discuss what you think are male and female attitudes to sex. Include in your discussion apparent contradictions such as a man should 'sow his wild oats', whereas a woman doing the same thing would be considered immoral by many people.
3. Some religions believe that sexual intercourse should be only for the purpose of having children. Argue the case for and against this belief. At the end of the discussion, how many members of the group agree with this view?
4. Girls start their periods at puberty – quite some time before they are emotionally ready to have a baby. As a group, check that you understand what happens during the menstrual cycle. Include such things as:

   - pre-menstrual tension
   - sanitary protection
   - old wives' tales
   - why periods vary from woman to woman
   - painful periods (dysmenorrhoea)

   After this, discuss why a girl of 13 or 14 is not yet ready to have a baby.

Infertility

It is very upsetting for a couple to find that they cannot have children. It is thought that about one in six couples may be infertile, and the number is steadily rising. In most cases, the couple do not realise that they have a problem until they try, unsuccessfully, to have a baby.

If a couple haven't conceived a baby after about one to two years, they may be offered various tests to find out the cause of their infertility.

Infertility can have many causes, and men and women are equally likely to be affected. In about a third of cases referred for treatment, the problem is a joint one.

Here are some of the common causes of infertility:

Female:

- damage to the uterus or Fallopian tubes
- no ovulation (no eggs produced)
- blocked Fallopian tubes

Male:

- no sperm produced
- too few sperm produced
- sperm which can't swim
- abnormal sperm
- blocked tubes which carry the sperm

The most common cause of female infertility is failure to ovulate, caused by hormone failure. Occasionally, a hormonal imbalance can make the vagina hostile to the sperm and prevent them from reaching the egg. In some women, the problem may be physical rather than hormonal. The hymen or tightness of the vagina may make penetration impossible. Infertility can occur at any time because of infections, diseases or growths. Severe stress can stop ovulation, or prevent the egg from travelling along the Fallopian tubes.

In men, infertility is mainly a result of sperm problems. There may be no sperm produced, or too few. The sperm may be abnormal, or unable to swim well enough to reach the egg. The tubes carrying the sperm may become blocked for various reasons, so that the seminal fluid would contain no sperm. Extreme stress may lead to impotence or premature ejaculation.

The doctor will refer the couple to the hospital or clinic in order for the cause of their problem to be diagnosed. The specialist will want detailed information about their medical histories, the woman's menstrual cycle, their sex life, and so on. The man will be asked to give a sperm sample into a sterile container. He should not have had sex for at least two days beforehand, to ensure that the sperm level is as high as possible. The sample will be checked to make sure there are enough healthy sperm. If all is well with the man's sperm, then tests will be carried out on the woman.

The woman will be asked to take her temperature regularly over a few months so that her ovulation pattern can be checked. A blood test to measure hormone levels may be taken about six days before her period is due. The specialist may also decide to examine a sample of her womb lining under a microscope to check whether she has ovulated. If the doctor suspects there may be physical abnormalities in the cervix, the couple will be given a post-coital test. Around the time of ovulation, they will be asked to have intercourse either the night before or on the morning of the test. The mucus around the cervix is collected and tested to see how the sperm are progressing.

To check for blocked Fallopian tubes, the doctor will inject dye into the Fallopian tubes through the cervix. An X-ray will show up any blockages. It is possible for the specialist to look at the uterus, Fallopian tubes, and ovaries through a laparoscope. The woman is given a general anaesthetic and a needle is passed through the abdominal wall. The abdominal cavity is filled with gas to distend it, and the laparoscope is passed through another small cut just below the navel.

Once treatment for infertility is started, it has been found that about one in twenty couples will conceive a baby naturally, probably because they feel less tense as their problem is being dealt with. If the problem is caused by the woman's failure to ovulate, then she may be given fertility-drug treatment. Very few courses of treatment do result in multiple pregnancies, although many couples find that they are expecting twins.

Blockages and other abnormalities in the man or the woman may be corrected by surgery, and infections can be treated with drugs. If the man's infertility cannot be treated, then the woman can become pregnant by artificial insemination, using either the husband's sperm if possible, or, if not, using sperm from a donor.

If the infertility is caused by emotional or psychological problems, the couple may be offered some form of psychotherapy or counselling.

In general, it is easier to treat female infertility than male infertility, but research is continuing all the time, and couples should be encouraged to remain hopeful about treatment. Infertility can be very damaging to a couple's sex life, as each month is spent wondering whether the woman has become pregnant this time. Couples in this situation need all the help and support they can get.

Assignments on infertility

1. Some of the help offered to infertile couples is controversial. Some people believe we should not meddle with nature or the laws of God by experimenting on live foetuses. As a group, discuss these controversial issues:

   - 'in vitro' fertilisation (test-tube babies)
   - artificial insemination by husband (AIH)
   - artificial insemination by donor (AID)
   - surrogate mothering

2. Couples who are trying unsuccessfully for a baby may suffer a great deal. Read this case study:

   > 'Jenny and her husband Tim had been trying for a baby for the past three years. Jenny felt very depressed, particularly every month when her period came, proving that, yet again, she was not pregnant.
   >
   > All around her, her friends and neighbours had young families. Jenny felt a failure as a woman; she felt useless. Tim tried to understand her feelings and to be sympathetic, but he too was full of self-doubt. He wondered why he couldn't perform the simple task of fathering a child.
   >
   > Jenny and Tim found that the fun had gone out of their sex life. Gradually, they found that "baby-making" had taken the place of "love-making", and this put stress on their relationship. Jenny found herself wondering who was to blame for the situation – herself or Tim. She felt they should discuss the problem together, but as time passed this became more difficult. Even visiting family and friends became a traumatic experience.
   >
   > People were beginning to ask when they intended having a family. "If only they wouldn't get at us," complained Jenny.'

   What advice would you give to Jenny and Tim? Separate your advice into:

a) practical help, i.e. what steps they could take to remedy the situation;
b) personal help, i.e. help them to come to terms with their situation and improve their relationship.

Useful addresses:

- National Association for the Childless, 318 Summer Lane, Birmingham B19 3RL
- Child (self-help charity), Farthings, Gaunt's Road, Pawlett, Near Bridgewater, Somerset
- British Pregnancy Advisory Service, Austy Manor, Wootton Wawen, Solihull, West Midlands B95 6BX
- National Marriage Guidance Council, Herbert Gray College, Little Church Street, Rugby, Warwickshire CV21 3AP
- Family Planning Association 27–35 Mortimer Street, London W1N 7RJ

Always enclose a stamped addressed envelope for your reply.

3. Find out what help is available in your area for childless couples.
   - Is there a certain length of time a couple have to have 'been trying' for a family?
   - Does the specialist see the couple together?
   - Is counselling help offered to couples with psychological problems?
   - Is there a long waiting list?
4. In pairs or individually, research into the causes and treatment of male and female infertility.

Pre-conceptual care in the community

'Pre-conception' is the term used to describe the time between a couple deciding that they would like a baby in the near future and when they actually conceive the baby.

For some time now, there has been an emphasis, quite rightly, on the need for good ante-natal care in the community. Posters and leaflets encourage the expectant mother to attend her ante-natal clinic for regular check-ups to ensure that both she and her developing baby remain healthy.

Recently, however, doctors and nurses have been realising the need for both parents to be healthy themselves before conceiving a baby. After all, a mother may have been pregnant for six to eight weeks before she is aware of it. By this time, it may be too late to prevent some of the problems that may develop.

Some health authorities now believe that pre-conceptual care is an important part of preventive medicine and have introduced pre-conceptual clinics, where parents can go for help and advice before starting their families.

Pre-conceptual clinics

These clinics are found either as part of the local maternity services or at one of the larger general practices. The clinics are usually held at a regular time, depending on demand.

The clinics cover two main areas:

1. They advise women who have had medical problems during previous pregnancies (such as one or more miscarriages, a difficult labour, birth defects in the baby, death of the baby at or near birth, pregnancy complications, or low-birthweight babies).

   They also help women who have existing medical problems (such as high blood pressure, diabetes, various blood disorders, heart disease, etc.).
2. They prepare women who are planning a pregnancy by checking that they and their partners are fit and not doing anything that could be harmful to the baby.

In most pre-conceptual clinics, the couple come about six months before they hope to conceive a baby. It is important that both the man and the woman attend because, although it is the woman who carries the baby, the man's health and lifestyle do affect his sperm. If a man is a heavy smoker or drinker, or has a certain disease, or works with dangerous drugs and chemicals, all these could have a harmful affect on his sperm.

Apart from these reasons, it is also important for the man to go with the woman so that he becomes closely involved in planning for the baby and doesn't see the part he plays as less valuable than his partner's.

At the clinic

The woman's blood is tested for rubella and anaemia, and she is checked for thrush and harmful sexually transmitted diseases (STDs), such as herpes, syphilis, etc.

If she has not had a cervical smear within the past three years, she will have one taken to check that she is healthy.

In addition to these checks, she is also given

advice on how to lead a healthy life. She is advised on:

- how to improve her diet
- how to give up smoking and alcohol
- how to keep to the correct weight
- how to lose weight if necessary
- the need to stop taking drugs that are not prescribed by the doctor
- the need to take light exercise daily
- the need to rest
- the need for the clinic to see both parents to avoid any health hazards which may be harmful to an unborn child

If she is taking the pill, the woman is asked to stop taking it three months before conception, so that her body has at least three 'natural' periods before conceiving.

If the doctor feels that the woman's diet is not rich enough in all the vital nutrients, she will be advised to take a course of tablets containing all the essential minerals needed for a healthy pregnancy.

Finally, if the couple have any other problems that may lead to stress or tension, they are encouraged to discuss these at the clinic, and any help or advice will be given.

Assignments on pre-conceptual care

Group work

Before carrying on with the assignment work, check that you understand the meanings of the following terms: rubella, anaemia, thrush, sexually transmitted diseases, cervical smear.

Discussion topics

1. Why do you think that the clinic would like the woman to have at least three 'natural' periods before conceiving a baby?
2. Why does it make good sense to improve the mother's diet before she becomes pregnant?
3. What type of jobs could be dangerous to a couple who are thinking of starting a family?
4. What are the benefits of closely involving the father in planning for a baby?
5. Discuss any problems a couple may have that could lead to stress or tension.

Group written assignments

1. Find out if there are any pre-conceptual clinics in your area. If there are, perhaps it would be possible to arrange for the doctor or nurse to come and talk to your group about the work done at the clinic. Write a letter of invitation.
2. If there is no clinic in your area, find out what is available for couples wanting advice before starting a family.

   Here are some ideas on where to write for information:

   - local GPs
   - family planning clinic
   - local maternity hospital
   - area health authority
   - Community Health Council
   - citizens advice bureau
   - Health Education Council

   (The addresses should all be in your local telephone directory.)

Individual work

Find out about the following:

1. Rubella:
   - What is rubella commonly called?
   - What effect can rubella have on the unborn baby (foetus)?
   - What measures exist to prevent the likelihood of pregnant women catching rubella?
2. STDs:
   - Find out the symptoms of thrush, genital herpes, and syphilis
   - How do they vary between men and women?
   - How could these diseases affect a pregnancy?
3. Anaemia:
   - What is anaemia?
   - List some of the symptoms of anaemia.
   - How can it be cured?
   - How can a healthy diet help to prevent anaemia?
   - Why is it important for the mother's blood to be healthy before she conceives a baby?
4. Smoking and alcohol:
   - Most people know that smoking and drinking are bad for a pregnant woman. Why do you think it is important to stop smoking and drinking before starting a family?

(Much of the information you need to help you with these assignments is available from the Health Education Council, which will be listed in your local directory. It produces many leaflets, booklets, slides, films, and packs which are free of charge.)

To sum up, the group can read the following case study about a young couple who are hoping to start a family in the near future. When you have read the case study and discussed some of the points raised at the end of the passage, there will be a chance for you to carry out some role play.

Case study

'Wendy and Alan had been thinking about starting a family. Last time she visited her doctor, Wendy had noticed a poster advertising the pre-conceptual clinic being held at the surgery. As they both thought it would be a good idea to check that they were healthy to be "on the safe side", Wendy decided to call in on her way home from work.

The doctor was very friendly and Wendy felt quite relaxed, as it was the same doctor that she saw for contraceptive advice.

The doctor asked Wendy plenty of medical details, such as her family history, her height, weight, menstrual cycle, whether she smoked or drank etc., but she also explained why these questions were necessary.

Wendy had never had a cervical smear, so she was advised to have one to check that she was healthy. At the same time, the doctor could check for any sexually transmitted diseases. This rather upset Wendy at first, but she soon realised that this was part of the general procedure.

As she couldn't remember whether she had had rubella as a child, Wendy was glad to hear that the blood test checked for that as well as anaemia.

While the doctor wrote more notes, Wendy had to pop out to give a urine sample so it could be examined for various disorders.

When she returned, they discussed diet and whether either she or Alan smoked or drank. The doctor was glad to hear that Wendy and Alan ate a low-fat, high-fibre diet, and felt that Wendy's weight was good. She was not so pleased to hear that they both smoked. Wendy said that she had intended giving up smoking when she became pregnant, but that Alan seemed unable to stop.

The doctor then suggested that they should both come to a talk she was giving later in the week about the importance of being fit and healthy before starting a family.'

Discussion points

1. What medical details did the doctor ask for, and why do you think they were important?
2. On the whole, Wendy and Alan seem to have a sensible attitude towards becoming parents.
   a) In groups, discuss whether this is always so, and what problems the doctor might meet with other couples.
   b) In your groups, re-enact some of these situations to the class.

3 Conception and pregnancy

Immediately after conception, most women do not realise they are pregnant. If the baby is planned, the mother will have made a note of the date of her last period, as this will help the doctor to work out the date the baby is due – the expected date of delivery (EDD).

It is only when she misses her period that the woman will realise she is pregnant. Two weeks after her period was due, she can go to her doctor's for a pregnancy test. Many women like to know sooner and will buy a home pregnancy testing kit from the chemist which may give the result only one week after the period was due.

A pregnancy starts as soon as the male sperm meets the female egg and fertilises it. The fertilised egg has genetic information from the father's sperm and the mother's egg in one cell. Within 48 hours, the cell starts to divide and subdivide. On about the 4th day, the egg starts to move down from the Fallopian tube to implant itself in the wall of the uterus, where the embryo will continue to develop (fig. 3.1).

On the 11th day, the *amniotic sac* develops to form a protective, fluid-filled covering around the embryo, and this is closely followed by the development of the *placenta*. The placenta has a very important role to play during the pregnancy. One end is attached to the wall of the uterus, and the other end is attached to the foetus by the umbilical cord. The placenta provides the foetus with nourishment from the mother, removes the foetus's waste products, and allows the foetus to 'breathe' by passing on oxygen from the mother's bloodstream.

The development of the foetus is covered in detail in many books, so rather than repeat what has already been adequately covered elsewhere, there is an opportunity for work in this area in the assignment section.

The pregnant woman

Pregnancy, which is generally assumed to last 40 weeks, is divided into three *trimesters*. The first trimester, from weeks 1 to 12, is often thought to be the most difficult, as the woman is adjusting to the physical and mental changes taking place in her body. She may well be suffering from one or more of the minor complaints of early pregnancy – morning sickness, heartburn, or varicose veins. Although these are called 'minor' complaints, they can be very depressing for the woman suffering them.

It is important that every expectant mother should have regular check-ups at an ante-natal clinic. As well as having the physical side of her pregnancy monitored, she will also be able to talk about any worries and problems she may have.

The father of the expected baby needs to be involved in the pregnancy. Nowadays, fathers play a greater role in parenthood than in the past, and ideally they should be involved from the start. The father's support and understanding, as well as helping the mother, may also be helping him to adapt to the changes he will need to make as a parent.

The second trimester, weeks 13 to 24, is usually considered to be the easiest. Most women have got over their morning sickness at this stage, and their bodies will have adapted to the hormonal changes that have taken place. During these weeks, the woman will begin to look pregnant and will need to wear maternity clothes.

Towards the eighteenth week, the woman may well feel the first movements of her baby. These foetal movements are called the *quickening*, and the woman can now begin to think of her baby as a living person.

The final trimester is from 25 weeks to birth. As the pregnant woman grows larger, she will become slower and more clumsy. She probably won't feel much like rushing around as she did in earlier preg-

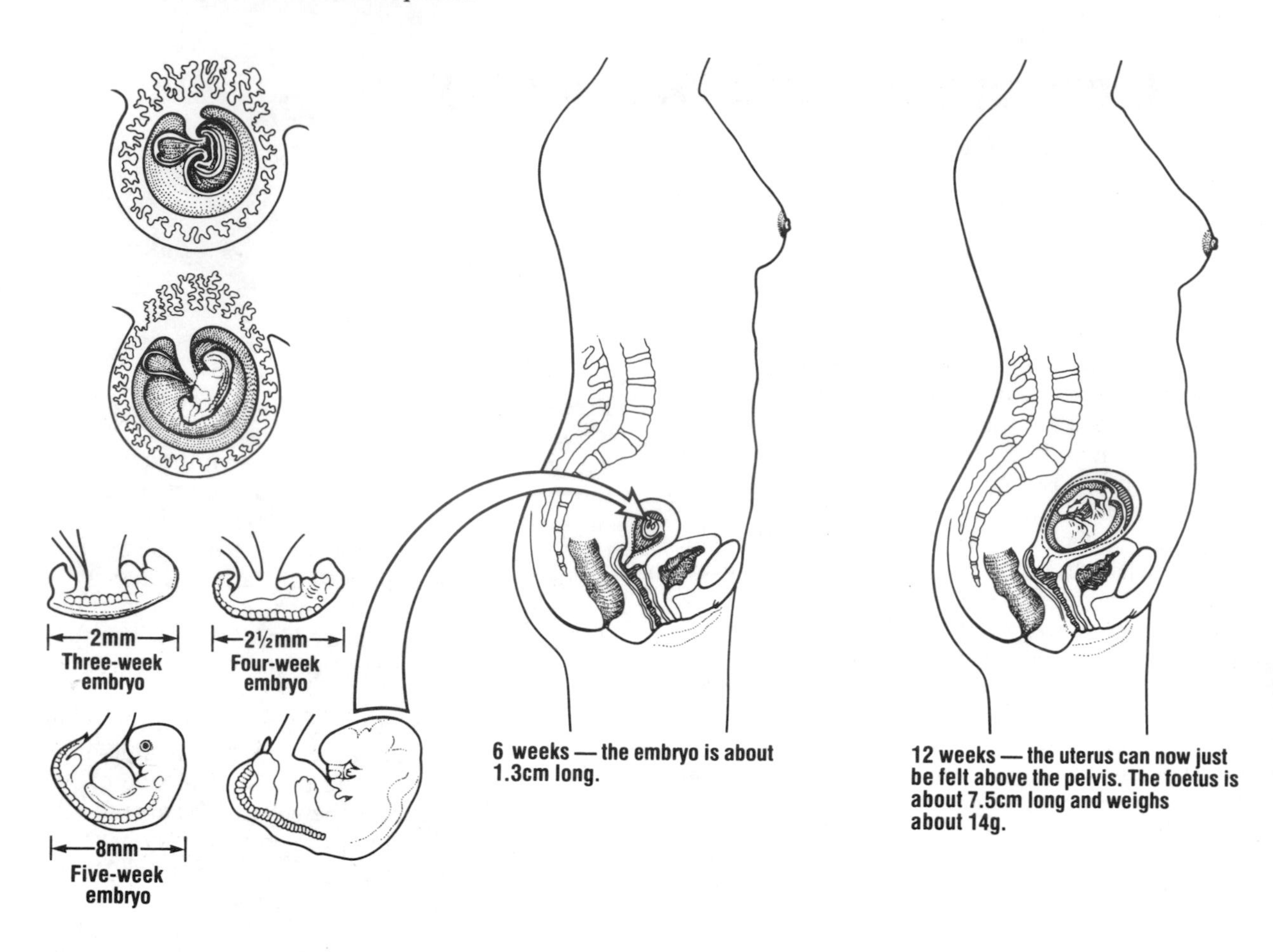

Fig. 3.1 The development of pregnancy

nancy, and she may be troubled by backache and sleeplessness. As the end of her pregnancy draws nearer, the woman prepares herself for the birth. She will feel both excitement and fear, in case anything should go wrong. Again, it is very helpful if she can talk over these worries with her partner, or at the ante-natal clinic.

Care of the expectant mother

It is very important that the pregnant woman should take good care of herself, because the healthier she is, the better able she is to cope with her pregnancy. The baby developing inside her will also benefit from a healthy and contented mother. Although ante-natal care is offered during her regular visits to the clinic, there is a great deal the pregnant woman can do herself.

Diet

It is always important to eat a well-balanced diet, but this is particularly necessary during pregnancy. A good diet will help keep mother and baby as healthy as possible, as well as keeping the woman's weight down and reducing the chances of her getting constipated. It is obvious that any pregnant woman will put on weight, but the extra weight should not be fat. Many women do put on extra fat during their pregnancy, perhaps because they mistakenly believe they should be 'eating for two', but sensible eating by avoiding too much fat and sugar can help keep weight down. Eating plenty of fibre will prevent constipation, and help to keep the calories down.

During pregnancy, the baby gets all its nutrients from the mother, so she needs to be sure to eat

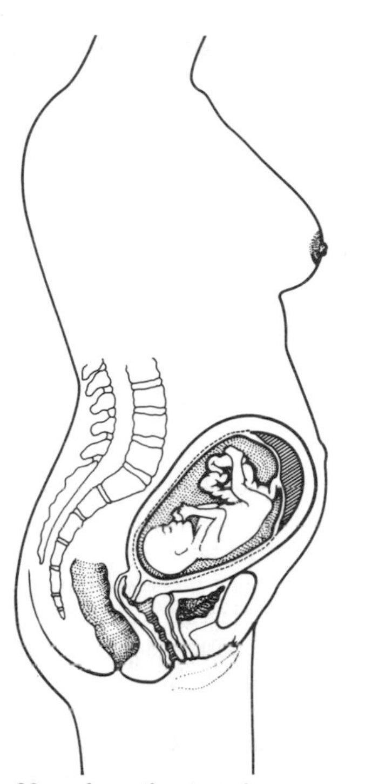

20 weeks — the uterus has now reached the level of the mother's navel and she becomes aware of the baby's movements. The foetus is now about 21cm long.

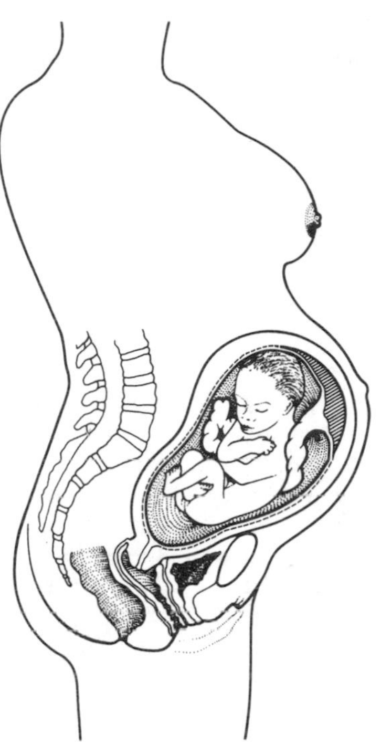

28 weeks — the uterus extends to about halfway between the navel and the sternum. The foetus is now about 37cm long and moves vigorously. It could survive if born at this stage.

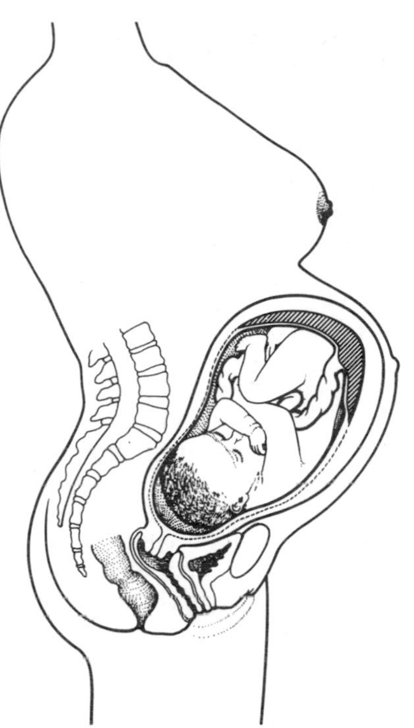

40 weeks — full term — the upper edge of the uterus moves down from high under the rib cage as the baby's head moves down into the mother's pelvis. (This is called 'engagement'.) Pressure on the mother's bladder increases, although her breathing and digestion become easier.

plenty of body-building foods that are rich in energy, high in fibre, low in fat, and full of all the vitamins and minerals she needs.

Exercise

It is a mistake to treat a healthy pregnant woman as if she were an invalid. Obviously, care needs to be taken over certain strenuous movements, but as long as the woman is well and has exercised regularly before her pregnancy, then there is no reason why she should stop. However, it is important that she should discuss this with her doctor early in her pregnancy and ask his or her opinion. A woman who has not taken regular exercise should not start strenuous exercise as soon as she finds herself pregnant, as the strain may cause unforeseen problems.

There are certain sports that should be given up during pregnancy, as they are dangerous and could be a hazard to the mother or foetus. Again, the doctor will give any advice. Exercises that are always recommended for pregnant women because they are safe for all – even those who do not exercise regularly – are walking, swimming, gentle yoga, and the exercises taught at ante-natal classes.

Sex

Unless her doctor tells her otherwise, a pregnant woman may carry on normal sexual relations with her partner. Many people believe that sex damages the foetus, but this is not so, as long as the mother is healthy. The couple who wish to continue their sex life will need to experiment with different positions as the woman's abdomen grows in size.

Pregnancy may affect the woman's feelings; she may be overtired and feel unnattractive, which will not increase her sexual appetite. Her partner may lose his interest in sex as the woman changes from a lover to an expectant mother. Many men find this change hard to come to terms with. The couple should try to discuss these problems before they feel that the pregnancy is coming between them.

Smoking, alcohol, and drugs

Smoking Ideally, any couple who are considering having a baby should both give up smoking before the baby is conceived. However, in many cases the woman will find she is pregnant while she and/or her partner still smoke. The man also needs to stop smoking because, while he continues to smoke, a non-smoker will breathe in his exhaled smoke and still suffer a health risk. This is called *passive smoking*.

Smoking affects the unborn baby by reducing the available amount of oxygen it needs to grow and develop. Research has found that babies born to smokers tend to be about 200 grams lighter than babies born to non-smokers. It is even thought that smokers' babies are more likely to be stillborn, more likely to die within a week of birth, and slower in their intellectual development.

Alcohol Drinking alcohol, like smoking, should ideally be stopped, or limited, by both partners before they start a family, as alcohol can cause genetic defects. Once she is pregnant, a woman who has only one or two drinks a day has a higher chance of a late miscarriage. Regular drinking throughout pregnancy can cause 'foetal alcohol syndrome', where the baby is born underweight, with slow mental and physical development, and occasionally physical abnormalities.

Drugs Drugs affect the unborn baby in many ways. Smoking marijuana affects the development of the foetus, and babies born to heroin addicts are addicted to heroin themselves, and need hospital treatment. The dangerous toxins in other drugs such as LSD, amphetamines, and solvents are thought to seriously affect the baby's development.

Almost all drugs may harm the foetus, especially in the first trimester. A pregnant woman should always ask the doctor's or pharmacist's advice before taking any drugs.

Today, the dangers of drinking alcohol, smoking, and taking drugs during pregnancy are well publicised. Perhaps the next step is to encourage couples to give up these drugs before they conceive a child.

The ante-natal clinic

Ante-natal care differs from area to area, so this section gives a very general account of the care offered. In the assignment section, there will be plenty of opportunities to find out what is available in your area.

Once her pregnancy has been confirmed, her GP will ask the woman about her medical history. She will be asked the date of her last period, so that she can be told the estimated date of her baby's birth. She will also be asked whether she has ever had an abortion or miscarriage, as this may affect her pregnancy. There will also be questions about any other medical conditions she may have that the doctor will need to keep an eye on.

She will be asked whether she wants to have her baby in hospital or at home – most women have already made that decision before they go to the doctor's. However, other factors may affect the decision. In a rural area, having a baby at home may be quite common, whereas in some cities and towns there may be no provision for home deliveries. In some areas, there may only be one hospital, so there is no choice. In the city a few miles away, there may be three maternity units to choose from. Whatever the choice, the woman should make the decision in consultation with her GP, who can explain the many factors involved.

When the woman is 10 to 12 weeks pregnant, she will be referred by her GP to the ante-natal or obstetric department of her local hospital. This visit is usually standard; after this, the woman must decide whether to carry on with her ante-natal care at the hospital clinic, or with her own GP, if this is possible.

One of the most common complaints about hospital ante-natal care is that it is unusual to see the same doctor twice, so it is impossible to build up a personal relationship. Many women prefer to have their check-ups with their own GPs where the staff and surroundings are familiar. However, this option is not available in all GPs' surgeries.

On her visit to the hospital, the woman is sent to the booking clinic, where a bed is booked for her in the labour ward. The questions she is asked are usually fairly standard; she is asked:

- to give a detailed medical and obstetric history and to say whether she has had an abortion, miscarriage, stillbirth, or previous babies, and whether she had any problems during previous pregnancies
- whether there is a family history of twins or multiple births
- her shoe size
- the sex and weight of any other children she may have had
- whether she has had any diseases such as diabetes, anaemia, high blood pressure, and whether she is taking any drugs

After the questions, the woman will be given some routine tests to make sure she is healthy and her pregnancy is progressing normally:

- She will be weighed, and her height measured.
- Her ankles and fingers will be checked for swelling (oedema).
- Her chest, heart, and blood pressure will be checked, and the results recorded.
- The size of her abdomen will be measured and noted (few doctors nowadays check this by carrying out an internal examination).
- Her urine will be tested.
- She will be given a blood test.

After this hospital visit, the woman will be asked to attend the ante-natal clinic either in the hospital or with her GP every four weeks until the 28th or 30th week of her pregnancy. At the hospital, she will be given a co-operation card (fig. 3.2). This card gives the woman's personal details, the name of her GP, the progress of her pregnancy, and previous medical and obstetric history. Inside the card is a chart to be completed with the details of the development of her pregnancy. It is filled in by the doctor or midwife each time she attends an ante-natal clinic. Later, details of the birth are filled in, and the mother is asked to keep the card as it gives a complete history of her pregnancy and birth. Should she decide to have another baby, the card will save her having to answer so many questions a second time.

After 28 to 30 weeks, the ante-natal visits become more frequent, and the woman is asked to attend every fortnight until she reaches 36 weeks, and then every week until the baby is born. For the mother who has decided to have most of her ante-natal care with her GP, there will probably be one more visit to the hospital towards the end of her pregnancy, but this varies from area to area.

Special tests

Apart from the routine tests carried out on every woman during her pregnancy, there are other tests that can tell a great deal about the developing foetus.

Ultrasound This, more commonly called a scan, uses sound-waves to make a picture of the baby in the womb (fig. 3.3). Many health authorities offer a scan to all pregnant women as a routine at around the 16th to 18th week of pregnancy in order to make sure that the baby is developing normally. If the mother is not sure of the date of her last period, the scanner can measure the size of the foetus so that the doctor can give an estimated date of delivery.

As well as being used as a routine check, the scan is valuable in detecting problems:

- It can detect certain abnormalities, such as spina bifida (a congenital malformation of the spinal cord) and hydrocephalus (a build-up of spinal fluid in the brain), and some congenital heart defects.
- It can be used to detect twins or multiple births, if these are suspected in later pregnancy. It can also show whether the baby is lying awkwardly in the uterus, which may cause problems at the birth.

Having a scan is a very simple procedure, and a routine scan is booked at the first hospital visit. The woman is asked not to pass water beforehand, and to drink a litre (two pints) of water two hours before the scan. This is because a full bladder will push the uterus up out of the pelvis and make it easier to see the foetus on the scan. For the scan to be taken, the woman is asked to lie on a high couch. Her abdomen is oiled so that the scanner can slide across it easily. The scanner is moved backwards and forwards over the abdomen, and a picture of the foetus will be built up on a monitor.

Although it is completely painless, many mothers worry that a scan may damage the unborn baby. So far, there is no evidence to suggest that scans are dangerous, but many hospitals carry them out only when strictly necessary.

Amniocentesis This is a test which involves removing some of the amniotic fluid surrounding

(3)

CONFINEMENT AND PUERPERIUM

CONFINEMENT   Date   Place

PUERPERIUM

MOTHER'S CONDITION ON DISCHARGE   Date

B.P.   Urine   Hb

Breasts   Nipples

Uterus

Perineum

Contraceptive Advice

BABY   Sex   Birth Weight

Condition at Birth

Date of discharge

Examination   Weight

Cong Malformation

Gestational Age

P.K.H

C.D.H·

Feeding

(4)

POST-NATAL EXAMINATION

DATE   Sig

Urine   B.P.   Hb

Symptoms and duration

Breasts and feeding

Abdomen

Pelvic examination

Rubella Vaccination

Treatment and advice

Cervical Smear   Date   /   /

SPECIAL NOTES

(Including recommendations for future pregnancies)

This card when completed should be returned to the family doctor

(1)

CO-OPERATION RECORD CARD FOR MATERNITY PATIENTS

NAME (Block capitals – surname first)

Address   Tel. No.

AGE   N.H.S. No.

FAMILY DOCTOR   Tel. No.

Address

DOCTOR BOOKED FOR MATERNITY SERVICES   Tel. No.

Address

MIDWIFE   Tel. No.

Address   ALT. No.

DOCTOR'S MATERNITY CLINIC   Tel. No.

HOSPITAL   OBSTETRICIAN

BOOKED FOR DELIVERY AT·   E.D.C.   /   /

RELEVANT FAMILY HISTORY   T.B.   Diabetes

High B.P.   Twins

PAST ILLNESSES, ALLERGIES, BLOOD TRANSFUSIONS, DRUG THERAPY, CONTRACEPTION

PREVIOUS OBSTETRIC HISTORY

No. confinements before 22 weeks..................................after 22 weeks.............

NORMAL MENSTRUAL CYCLE   D.L.M.P. (First day)   E.D.C.

Oral Contraception   YES/NO

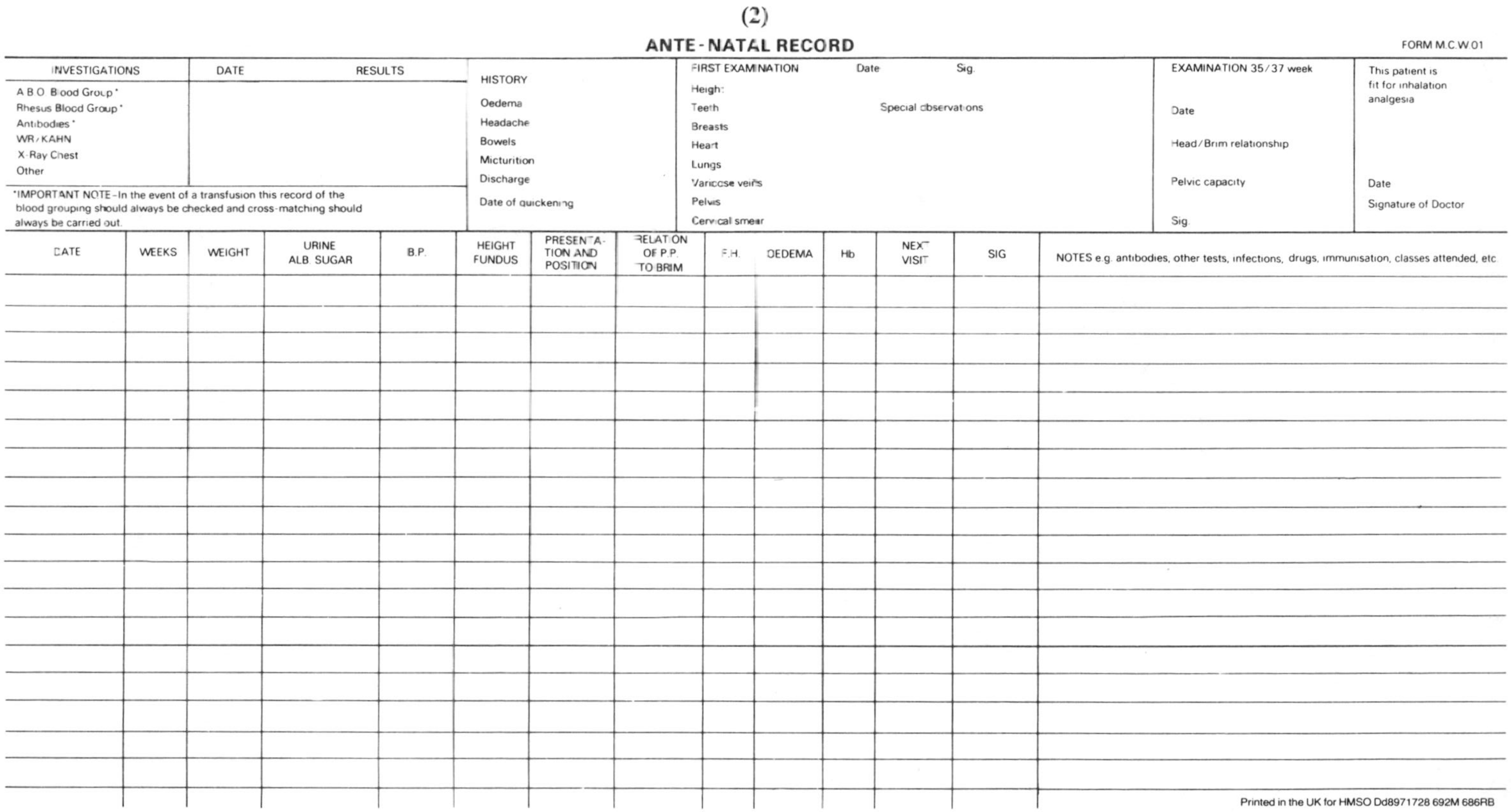

(2)

ANTE-NATAL RECORD

FORM M.C.W.01

INVESTIGATIONS	DATE	RESULTS	HISTORY	FIRST EXAMINATION Date Sig.	EXAMINATION 35/37 week	This patient is fit for inhalation analgesia
A B O Blood Group* Rhesus Blood Group* Antibodies* WR/KAHN X-Ray Chest Other			Oedema Headache Bowels Micturition Discharge Date of quickening	Height Teeth Breasts Heart Lungs Varicose veins Pelvis Cervical smear Special observations	Date Head/Brim relationship Pelvic capacity Sig.	Date Signature of Doctor

*IMPORTANT NOTE–In the event of a transfusion this record of the blood grouping should always be checked and cross-matching should always be carried out.

DATE	WEEKS	WEIGHT	URINE ALB. SUGAR	B.P.	HEIGHT FUNDUS	PRESENTA-TION AND POSITION	RELATION OF P.P. TO BRIM	F.H.	OEDEMA	Hb	NEXT VISIT	SIG	NOTES e.g. antibodies, other tests, infections, drugs, immunisation, classes attended, etc

Printed in the UK for HMSO Dd8971728 692M 686RB

Fig. 3.2 A co-op card

Part 1, the first page of the folded card, shows obvious information such as name and address, for administrative purposes, but also records relevant medical information and details of previous pregnancies ('previous obstetric history'). 'DLMP' means 'date of the last menstrual period', and 'EDC' is the 'expected date of confinement'.

Inside the card (2) is the complete ante-natal record showing (from left to right) the date of the ante-natal examination, the number of weeks the pregnancy has reached, the woman's weight, details of the urine test, her blood pressure, the height of the fundus (the top of the uterus), the position of the foetus, the relationship of the foetus's head to the brim of the pelvis, whether the foetal heart can be heard, whether the woman has oedema (swelling of fingers, ankles, etc.), the haemoglobin level in her blood, and the date of her next visit. Any other relevant comments can be made in the 'Notes' column.

Part 3 deals with the birth of the baby (confinement) and the time immediately after the birth (puerperium). The mother's state of health is noted before she leaves the hospital, and details of the baby's health are also checked and recorded.

Part 4 is completed at the woman's post-natal check-up, about 6 weeks after the birth.

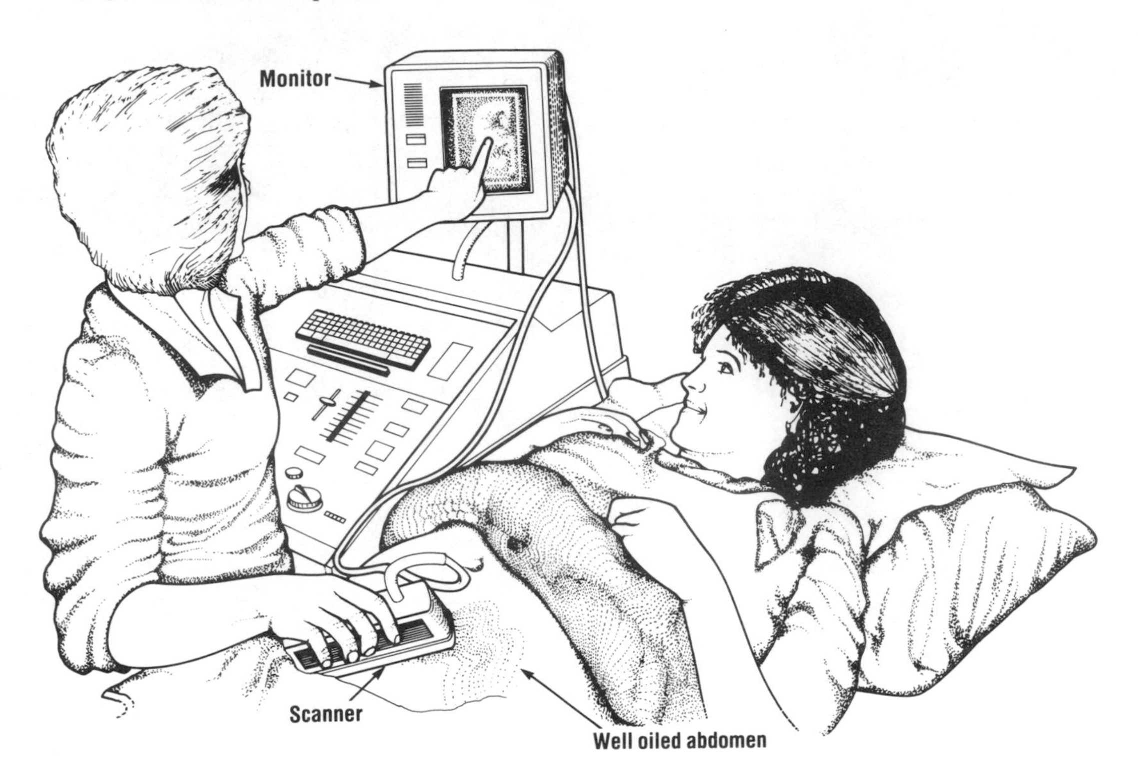

Fig. 3.3 Ultrasound

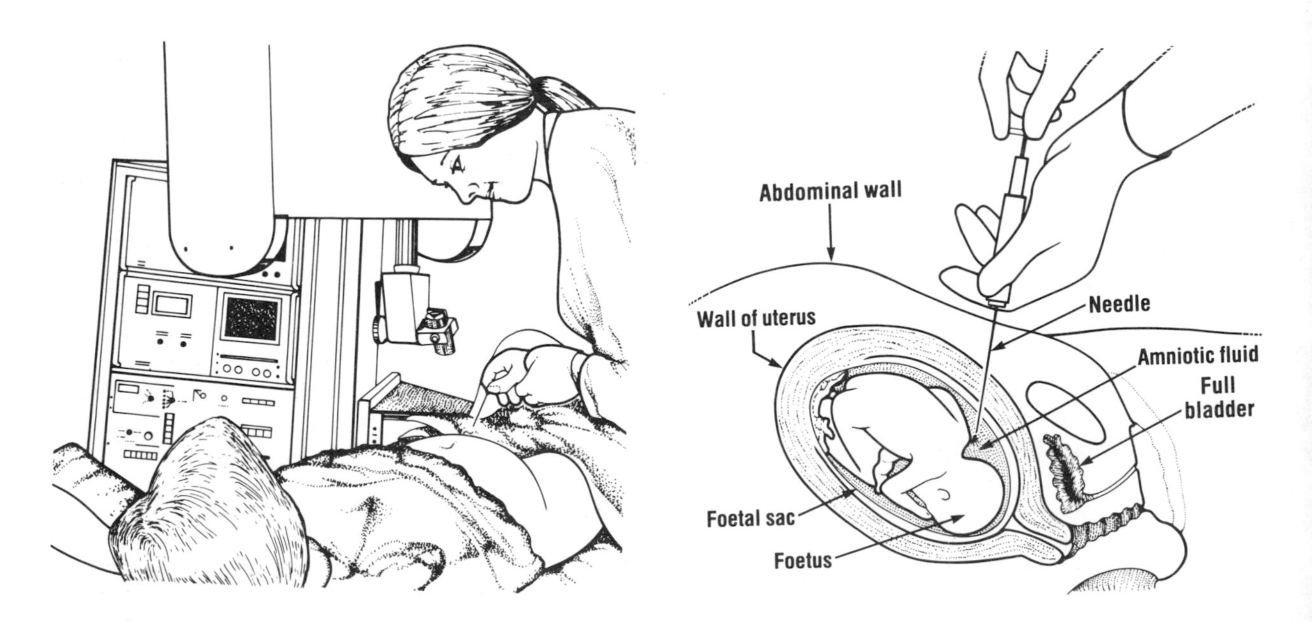

Fig. 3.4 Amniocentesis

the foetus in order to check whether or not there are any abnormalities (fig. 3.4). It is carried out on women only where there is thought to be a risk, i.e.

- older women (over 38), as there is a greater likelihood of having a Down's syndrome baby
- if there is a family history of spina bifida, haemophilia, muscular dystrophy, or Down's syndrome

After an amniocentesis, the woman has to wait two to three weeks for the results, and this long wait can be very distressing. There is also the slight risk that the test may cause a miscarriage, or damage the foetus. These dangers, although unlikely, coupled with the high cost of the test, mean that amniocentesis tests are not carried out unless absolutely necessary.

To have the test, the pregnant woman first has an ultrasound test, which will show the doctor where the placenta and foetus are in the woman's uterus. After this, the skin around the abdomen is cleaned, and a local anaesthetic is given to prevent any pain. A long hollow needle is then passed through the abdominal wall, and 5–10 ml of amniotic fluid are drawn off into a syringe. The fluid is then tested in the laboratory to see if there are any abnormalities.

Rhesus factor Rhesus factor is a problem that will affect a second pregnancy if the mother's blood is Rhesus negative and the father's blood is Rhesus positive. The condition means that the woman will produce antibodies in her bloodstream. Although these antibodies will not affect the first baby, there will be complications in any future pregnancies. It is very important to treat this condition, so the mother is given an injection within three days of the expected date of delivery.

Ante-natal classes

Most doctors and midwives would recommend that the expectant mother attends ante-natal classes, which are there to prepare the prospective parents for the birth of their baby.

Classes may be run by the hospital, local health centre, or associations such as the National Childbirth Trust. The classes vary, but broadly speaking they aim to cover topics such as keeping healthy during pregnancy, what to expect in labour, relaxation techniques, and baby-care.

These classes are particularly useful for the first-time parent, as they offer an opportunity to:

- see where the baby will be born
- ask any questions and discuss any problems
- meet other parents in the same situation

Although the classes are mainly geared towards the mother, many offer a chance for fathers to become involved.

Minor complications of pregnancy

Even a woman who considers herself to have had a trouble-free pregnancy may well have suffered from at least two of the problems common in pregnancy. It is important that women know about these minor health problems, so that they can recognise when they should ask for medical help.

Backache

This is particularly common, mainly because the woman has to adapt to the changes caused by her growing abdomen. Another cause is that, during pregnancy, some of the ligaments which support the spine become softer in preparation for the birth. Many of these problems can be reduced if the woman is careful to lift properly.

Morning sickness

This is caused by the hormone changes in early pregnancy. Some women may feel slightly nauseous, whereas others may actually be sick. Morning sickness tends to go by the 12th to 14th week of pregnancy. Many women find it helps if they eat little and often, and cut out greasy and rich foods.

Constipation

Again, this is a complaint caused by hormonal changes. If left untreated, constipation can cause haemorrhoids, or piles (varicose veins in the rectum). The best way of treating constipation is to eat a diet rich in fibre rather than taking laxatives. A pregnant woman should never take a laxative without first asking her doctor, midwife, or pharmacist's advice. Some laxatives may be too strong to be taken safely during pregnancy.

Frequency

Many expectant women find that one of the first signs of their pregnancy is the need to pass water more frequently. For some women this only lasts a few weeks, yet others may have frequency until their baby is born. Doctors are not sure what causes this frequency in early pregnancy, but in later pregnancy it is caused by the baby pressing on the bladder.

Indigestion
Indigestion, or heartburn, affects most women at some time during their pregnancy, even if they have had no previous experience of this. Heartburn is particularly uncomfortable, and tends to happen when the woman lies down or stoops. There is not a great deal that can be done to help, but she could try taking some of the gentle antacids recommended for pregnant women.

Haemorrhoids
Haemorrhoids, or piles, may be caused by constipation, or by pressure from the baby's head pushing down on the back passage. Piles can be very uncomfortable, and even painful, so should always be mentioned to the doctor or midwife, as there are simple remedies available.

Stretch marks
Stretch marks usually affect the skin on the abdomen and/or the breasts. They are nothing to worry about, but many women think them unsightly. There are creams on the market which promise to help prevent stretch marks, but they are not effective because the marks are well below the skin's surface. Stretch marks gradually fade after the birth, but never completely disappear, and remain as silvery lines. There is no need for a mother to be ashamed of these marks, as they are the result of a natural process.

Swollen feet and ankles
These occur fairly frequently in pregnancy because the body tends to retain water, especially on a hot or tiring day. However, this condition should always be mentioned at the clinic, as it can be a sign of pre-eclampsia, which can be very dangerous.

Major complications of pregnancy

As we have already discussed, one of the main reasons for going to the ante-natal clinic is to check the mother's health and the development of the foetus, in order to make an early diagnosis of any complications. Although the majority of pregnancies are straightforward, with maybe only a few minor ailments, sometimes there are major complications.

Miscarriage
One of the first complications which may arise is a miscarriage. This is often called a 'spontaneous abortion', and should not be confused with a clinical abortion, used to terminate a pregnancy.

Most miscarriages happen within the first 14 weeks of pregnancy, although technically they can occur at any time up to 28 weeks.

The first signs of miscarriage are:

- bleeding from the vagina
- abdominal pains

If a pregnant woman develops either or both of these symptoms, she should contact her doctor straight away. The doctor will probably recommend bedrest. Sometimes the pregnancy can be saved this way, but at other times the miscarriage may be inevitable.

The most common cause of early miscarriages is abnormality of the foetus. Later in the pregnancy, miscarriages may be caused by what is called an *incompetent cervix*, where the cervix is unable to keep tightly closed during pregnancy.

After a miscarriage, the woman may need a small operation called a *dilatation and curettage* (commonly known as a 'D & C'), which will clean out the uterus and remove the remnants of the pregnancy. Her doctor will tell her when to try for another baby.

If the woman has had more than one miscarriage, she may need medical help in order to have a successful pregnancy.

Ectopic pregnancy
A rare complication that can occur in pregnancy is called an *ectopic pregnancy*, which is when the foetus develops outside the uterus, usually in the Fallopian tubes.

In an ectopic pregnancy, the woman will miss her period as in a normal pregnancy, but then she will have increasingly bad abdominal pain and vaginal bleeding.

One of the causes of a Fallopian ectopic pregnancy is damaged Fallopian tubes, where the tubes are unable to pass the fertilised egg along to the uterus.

An ectopic pregnancy has to be removed by an operation.

Foetal death and stillbirth
The other major complications – pre-eclampsia, eclampsia, and antepartum haemorrhage – are

covered in the assignment section. However, perhaps the worst complication in a woman's pregnancy is the death of her baby in the uterus or at birth.

Nowadays, many of the causes of stillbirth, such as pre-eclampsia, eclampsia, diabetes, Rhesus incompatability, and so on, which caused so many deaths in the past, can be treated during pregnancy. However, in the United Kingdom every year about 6000 babies die before birth, and the same number soon after birth.

One of the main causes of foetal death is strangulation, caused by the umbilical cord becoming wrapped around the baby's neck. This can be checked by foetal monitoring during labour, so that the baby's life can be saved by an emergency Caesarian-section operation.

If a pregnant woman has not felt her baby move for more than 24 hours, she should contact her midwife or doctor. If the baby has died in the womb, the woman will be admitted to hospital for an induced labour. However, in many cases of foetal death, labour will follow quite soon.

It is obviously very upsetting for the parents to cope with a stillbirth. As the pregnancy progressed, the mother will have become more attached to the baby inside her. The couple may have planned ahead and bought some baby clothes, a cot and so on, and are now faced with bereavement rather than having a birth to celebrate.

It has been found that parents come to terms with their loss better if they are allowed to hold the dead baby, and maybe even have a photograph to keep. If they wish, the baby should be named and given a funeral, and there should be an opportunity for the parents to talk about their experience with someone qualified to help them.

A post-mortem will be carried out on the dead baby, and a consultant will explain the results of the examination to the parents. However, there is not always an explanation, so the couple need reassurance that it is unlikely to happen again.

Rights and benefits

There are many rights and benefits available to pregnant women, depending on their individual circumstances. The system is very complicated, and any woman having difficulty working out what she can claim should contact her local Department of Health and Social Security (DHSS) or citizens advice bureau. If this is not possible, then her doctor, midwife, or health visitor will ask a social worker to help her to sort out her claim.

The various grants and allowances are not discussed in this section as they are subject to change. Instead, they will be investigated in the assignments.

Assignments on conception and pregnancy

1. The placenta plays an important role in pregnancy. It:

   - nourishes the foetus
   - removes foetal waste products
   - carries oxygen to the foetus
   - secretes hormones to keep the pregnancy stable
   - carries the mother's antibodies to the foetus
   - creates a barrier to prevent most diseases reaching the foetus

   Find out about the placenta and the umbilical cord. Include any problems arising from placental abnormalities such as:

   - placenta praevia
   - placental haemorrhage
   - detached placenta
   - inadequate placenta

2. Find out how the foetus develops during pregnancy. There are plenty of books available, with good diagrams and photographs. The BBC's 'Horizon' programme called 'Miracle of life' shows conception and pregnancy, and is particularly relevant.
3. A great deal is known nowadays about the dangers of smoking, drinking, and drug-taking during pregnancy.

   a) In pairs, write some guidelines for a woman who wants to give up smoking before conceiving a baby. Include:

      - the dangers to her health
      - the dangers to the baby's health
      - the dangers of passive smoking
      - ways of giving up smoking, including help available

   b) In the same pairs, find out the dangers of drinking during pregnancy. Include:

      - how heavy drinking may affect the man's sperm

- the dangers to the developing foetus
- foetal alcohol syndrome

4. Problems can arise during pregnancy, and these may be minor and fairly common, or more dangerous. Look at some of these problems in more detail:

- morning sickness
- heartburn
- miscarriage
- ectopic pregnancy
- pre-eclampsia and eclampsia
- antepartum haemorrhage
- high blood pressure (hypertension)
- diabetes
- intrauterine death and stillbirth

Find out, if possible, the symptoms and causes, and whether there is any treatment available.

5. A good diet during pregnancy is essential for many reasons:

a) To prevent the mother gaining too much weight:

- Suggest ways in which a pregnant woman can reduce her fat intake.

b) To prevent constipation:

- Suggest meals rich in fibre that will help prevent constipation, as well as costing less.

c) To give her baby all the essential nutrients, the woman needs to eat body-building foods that are high in energy and fibre and low in fat. She also needs her daily vitamin allowance. Her calorie requirements will increase as her pregnancy develops.

- Compare the diet of a woman before and during her pregnancy.

6. The saying 'She's eating for two' is often heard. As a group, discuss how true this is.
7. A pregnant woman may have to decide where to have her baby. Find out what choice she has in your area.
8. Find out which of these ante-natal schemes are available in your area:

- shared care, where the GP gives ante-natal check-ups at his or her surgery with back-up from the hospital
- GP unit
- Domino scheme
- home delivery
- hospital ante-natal and delivery

List the advantages and disadvantages of each of these schemes.

Useful contact points:

- Head of Midwifery Services in your local health district
- National Childbirth Trust, 9 Queensborough Terrace, London W2 3TB
- Community Health Council (address in the local telephone directory)

9. At her first hospital ante-natal clinic visit, the woman will be asked many questions, and tests will be carried out. All this may take a long time, and there is usually a wait, but it is essential to check her health and her baby's development in order to reduce the likelihood of problems. One way we can understand the importance of these tests is to find out why and how they are carried out.

a) In this section of the book we have used these terms: obstetrics/obstetrician, midwife, health visitor, general practitioner (GP). Find out what these people do, and write a brief summary of how they help a pregnant woman.

b) As a group, discuss why you think the hospital asks these questions:

- whether the mother has had an abortion or stillbirth
- whether she has had a baby previously, and if the pregnancy and birth were normal
- whether there is a family history of twins or multiple births
- the size of her feet
- whether she has any diseases such as diabetes and anaemia
- whether she is taking any drugs

If there were any questions you could not answer, check that you now understand.

c) Discuss some of the tests that are carried out:

- Why is the mother weighed?
- Why is her height measured?
- What is oedema, and why is it dangerous?

- What are the problems of varicose veins in pregnancy?
- Why are heart, lungs, and blood pressure checked?

d) Blood and urine tests are particularly important, as many of the problems that occur during pregnancy can be detected by testing the blood and urine. In a normal pregnancy, the mother will have two blood tests, but her urine will be tested at every visit. Find out about the blood and urine tests:

- What are they?
- How are they taken?
- How are they tested?
- What may the tests show?

10. Find out the pattern of ante-natal care followed by your own GP. These questions may help you:

- Does your GP offer ante-natal care?
- If so, is there a separate clinic, or does the expectant mother attend normal surgery hours?
- If not, which hospital is the woman referred to?
- Does the midwife or health visitor attend the ante-natal clinic regularly to offer help and advice to the pregnant woman?
- Is the woman offered a choice of where she has her baby?
- Can she have her baby at home if she wishes?

Think of any other useful and relevant questions a newly pregnant woman may need to ask. Write the questions down and check with the receptionist at your GP's surgery whether someone there would mind helping you with your research.

11. Talk to friends or relatives who have had babies in the last ten years. Ask them for their opinions of the ante-natal care they received. Compare the more recent accounts with those of some years ago.

- What were the most frequent criticisms of the mothers who had their babies some years ago?
- What were the criticisms of the more recent mothers?
- Do these accounts show any improvement in ante-natal care over the past few years?

12. Amniocentesis is not carried out routinely because of the possible dangers. As a group, discuss any of the following issues:

- Should amniocentesis be carried out at all?
- If the results show an abnormality in the foetus, what problems and decisions face the parents?
- Should amniocentesis be more routinely available to help reduce the number of handicapped babies?

13. Look into the problem of the Rhesus factor: it is well documented in books on pregnancy.
14. Rubella, or German measles, is not a dangerous disease in itself. However, if a pregnant woman develops rubella in the first 12 weeks of her pregnancy, there is a risk that her baby may be born handicapped. Nowadays, girls are given a vaccination to prevent this from happening, but many pregnant women are too old to have had this routine vaccination, so the blood of all pregnant women is tested for rubella antibodies.

- Find out about the dangers to the foetus if the mother catches rubella.
- Do you remember having the vaccine? (girls only)
- How well-publicised is the vaccine in your area?

15. a) Find out about the ante-natal classes in your area.

- What are they called?
- Where are they held and when? Is there a crèche available?
- Are they free?
- Who runs them?
- Are fathers encouraged to attend?
- What do the classes offer?

b) Many ante-natal classes offer breathing and relaxation exercises to help the mother cope with her labour. Find out about the relaxation and breathing classes offered in your area. Who runs the classes, what do the mothers do, and what are the benefits of these classes?

16. Many of the minor problems occurring in pregnancy can be helped or treated in some way. Give advice to women suffering from these conditions:

- backache
- morning sickness
- some suggestions to help avoid indigestion (remember it is up to the doctor to recommend any medication)
- some simple hints to help overcome swollen feet and ankles (if, after visiting the doctor, the woman has been told it is not pre-eclampsia)

17. Miscarriage and stillbirth are very distressing whenever they occur in a pregnancy.

    a) In pairs, find out this information:

    - Medically, what is the difference between miscarriage and stillbirth?
    - What are the signs and symptoms of each?
    - What should a woman do if she is showing any of these signs?
    - Explain any treatment the doctor or midwife might give.

    b) Discuss how parents may come to terms with a stillbirth. Useful addresses:

    - Stillbirth and Neonatal Death Society (SANDS), Argyle House, 29–31 Euston Road, London NW1 2SD
    - Miscarriage Association, 18 Stoneybrooke Close, West Bretton, Wakefield WF4 4TP

    If you write to either of these groups, send a stamped addressed envelope and explain why you would like the information.

    - The Health Education Council has a leaflet, M20, on coping with the loss of a baby.

    c) Find out if there is any local support group in your area for parents who have lost a baby.

18. There are many rights and benefits available for pregnant women and parents. One of the major problems is finding out which can be claimed. Here is a list of some rights and benefits available:

    - maternity grant
    - maternity allowance
    - maternity leave
    - maternity pay
    - time off for ante-natal clinics
    - free prescriptions and NHS dental treatment
    - free milk and vitamins
    - supplementary benefit
    - child benefit
    - one-parent benefit
    - family income supplement

    In pairs, divide up the list and find out about each of the rights and benefits. Include

    a) what the right or benefit is,
    b) who is eligible to claim it,
    c) when the claim should be made,
    d) how to claim.

    You may find the following helpful:

    - a social worker may be willing to come and talk to your group about claims (contact him or her at the social-services department, listed in the telephone directory)
    - citizens advice bureau
    - social-security office
    - leaflets produced by the DHSS often found at the post office, DHSS offices, doctors' surgeries, and chemists'

4 Labour and birth

Preparing for the birth

Towards the end of her pregnancy, the woman will feel a variety of different physical sensations.

Because of the size of her abdomen, she may feel uncomfortable and find it difficult to sleep at night. Any exercise will leave her short of breath until about the 36th week of pregnancy, when the baby's head will descend into the pelvis. This is called the 'lightening', and it describes the relief she will feel as she is able to bend down again. (In a second or subsequent pregnancy, the baby's head may not descend until just before labour begins, as the mother's uterus is big enough to take the full-term baby.)

The pressure of the baby's head on her bladder will make the woman need to make frequent trips to the toilet for the last few weeks of her pregnancy.

As well as the physical changes, the woman's emotions may be mixed too. She may feel fear that all will not be well with the baby, coupled with impatience over what seems like an endless pregnancy. She may also feel guilty because she is no longer sexually interested in her partner, as all her thoughts are directed towards becoming a mother. However, many women feel particularly serene during their pregnancies, and enjoy every moment. Everyone is different, and the same applies to pregnancy – many women say that each separate pregnancy is different, and one may be more enjoyable than another.

Preparing for the birth takes a certain amount of planning. As well as taking the practical steps such as buying equipment and so on, the expectant mother needs to prepare her body for childbirth.

Relaxation and breathing exercises are a good way of preparing the muscles for birth. Sometimes these exercises make it possible for the woman to give birth without using pain-relief methods, but it is important that she is not made to feel guilty if she does need pain-killers.

The woman will have been mentally preparing herself for motherhood for some time now. Many women find that towards the end of their pregnancies they are quite happy to become less sociable and relax in the security of their homes.

There is plenty of practical work to be done before the baby is born, and this is one way in which the future father can become involved.

Equipment

It is a good idea for the couple to sit down and think about what they will need after the baby is born. It is too late to leave it until after the birth, because the mother will be very busy with the baby, and probably won't have the time to think about shopping. To help them decide just what they will require, it is a good idea to think under the headings of food, warmth, and washing, as these are the baby's three main areas of need.

The father's role

During a pregnancy, most of the action is centred around the woman and, in many cases, the man's role is forgotten in all the excitement.

It is important that the future father should be involved in the pregnancy, and not treated as a bystander. In many areas, men are encouraged to come to ante-natal classes to find out about pregnancy, birth, and baby-care, and they are also shown how to help their partners during labour.

Sadly, some men become jealous that their partners are no longer giving them as much attention as they did before the pregnancy. Other men find it difficult to accept the changes in the woman's body. Perhaps nowadays these cases will become more rare, as pregnancy becomes more of a 'joint effort', with the man being encouraged to play a positive role in preparing for the birth of his baby. The help and support a man gives his partner during pregnancy can strengthen the couple's relationship in the future.

Labour and birth

Towards the end of her pregnancy, the woman will be advised to have a suitcase ready packed for the hospital, if she is having her baby there. She is given a list of what to pack at the ante-natal clinic.

If she has decided to have a home delivery, her midwife will have told her what to have ready at least one month before the baby is due. The expected date of delivery given by the doctor or clinic is only a rough guide, and the baby probably will not arrive on that exact date.

Many women pregnant for the first time worry that their labour symptoms will be a false alarm. Towards the end of her pregnancy, a woman will feel quite strong, but painless, contractions which are called *Braxton Hicks* contractions. These are quite often mistaken for labour pains, but the midwife would rather the woman contacted her, even if it does turn out to be a false alarm.

Labour starts in one of three ways:

1. *Contractions of the uterus* These usually start as pains in the lower back, and occur about every 15 to 20 minutes. Later – sometimes quite a few hours later – the contractions become stronger and more frequent.
2. *A show of mucus* Some women have what is called a 'show', which is a pinkish discharge from the vagina. This is the plug of mucus that has sealed the cervical canal during pregnancy, as it becomes dislodged by labour. There should not be a lot of bleeding with this, so any heavy bleeding should be reported immediately, as it may be a sign of haemorrhage from the placenta.
3. *The breaking of the waters* This happens when the amniotic sac surrounding the baby breaks naturally, and the amniotic fluid leaks out of the vagina. This may break at any stage during labour and, contrary to any old wives' tales that we hear, an early break does not cause a long and painful 'dry' labour.

It is often difficult for the woman to decide when to call the midwife or the hospital. Generally, the first stage of labour will last longer for a woman expecting her first baby. A woman expecting her second or subsequent baby may not have long to wait before her baby is born. Most hospitals recommend that the woman should contact them when the contractions occur every 10 minutes.

The process of labour is divided into three stages – the first, second, and third stages.

The first stage

The first stage starts in one of the three ways already mentioned. The uterine contractions gradually widen the cervix: this process is called *dilatation* (fig. 4.1). By the time the baby is ready to be born, the cervix will have dilated to 10 cm.

The cervix takes quite some time to dilate the first 3cm but, after that, the speed of dilatation then increases, and the expectant mother should be in hospital at this stage. The contractions will be quite strong and painful, coming about every five minutes, and eventually every two.

By the time the woman gets to the hospital, the midwife will have her pregnancy notes to hand and will ask her some questions about the progress of her labour. The woman will then be asked to undress and put on a hospital gown.

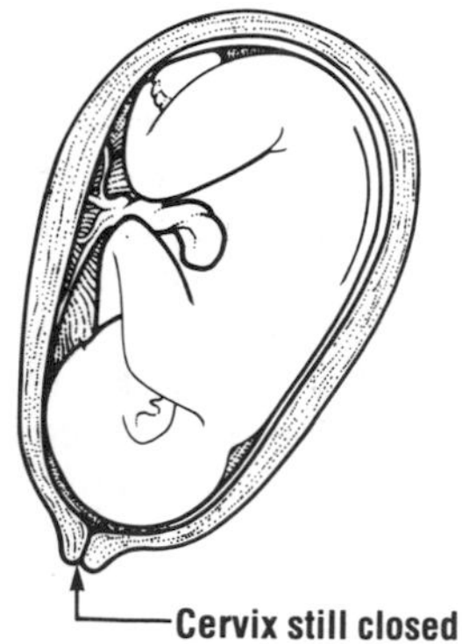

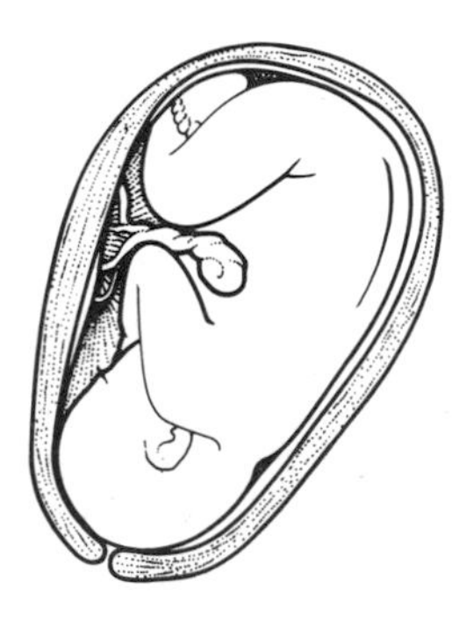

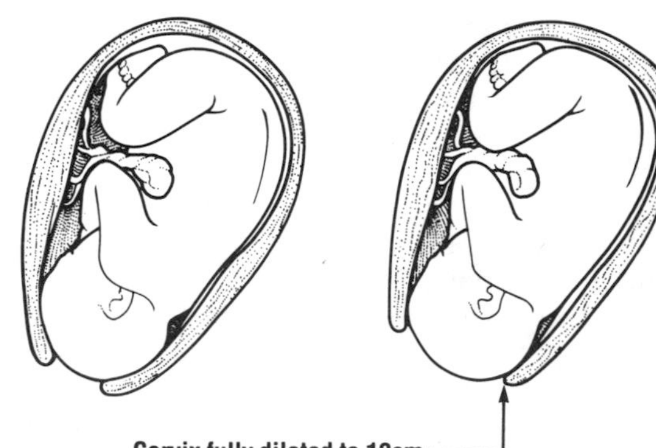

Fig. 4.1 The dilating cervix

After the midwife has checked the baby's position and heartbeat, she will take the woman's pulse, blood pressure, and temperature, and give her an internal examination to see how far her cervix has dilated.

In the past, women had their pubic hair shaved, but fortunately nowadays this practice has stopped. The woman will also be asked whether she has opened her bowels recently and, if not, she will be offered an enema or suppository.

What happens next varies from hospital to hospital. Many hospitals encourage an active birth, and the woman moves around as much as possible in order to relieve the pain of the contractions. Other hospitals may take the more traditional approach, where the woman is taken to a ward or room for the first stage of labour. If the labour is to be electronically monitored, she will need to lie in bed, and this will restrict her movement.

Occasionally, the woman may need to have a drip put into a vein. The drip may contain either glucose, to increase her energy, or a hormone to speed up a slow labour.

During ante-natal classes, the pain relief available will have been explained to the woman. There is always some pain during labour, but the level of pain varies from person to person.

There are several ways of relieving pain during labour, and the woman will be asked to decide which method she prefers. Many women decide that they would like a 'natural' childbirth, using no pain-relieving drugs at all. In these cases, they may use relaxation and breathing techniques to help them.

The other methods of pain relief are as follows.

Pethidine This is a pain-relieving drug that is given by injection. It takes about 20 minutes to take effect, and lasts for two to four hours. As it is a strong drug, it should not be given too close to delivery of the baby, as it can affect the baby's breathing. One disadvantage is that it may leave the woman less aware of what is going on around her, and therefore less able to cope with her contractions, and this may prolong the labour. A side-effect of pethidine is that the woman may feel, or be, sick.

Epidural anaesthesia An epidural is an anaesthetic injected into the spine which cuts off feeling below the waist (fig. 4.2). It must be set up by an anaesthetist, and this takes 10 to 20 minutes.

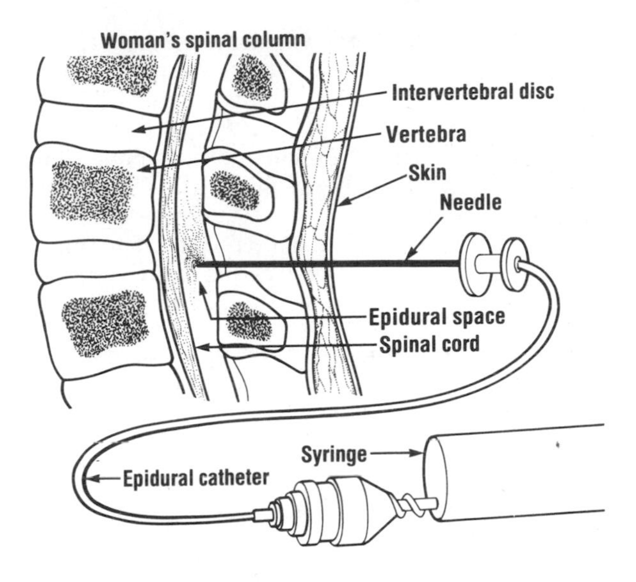

Fig. 4.2 Epidural anaesthesia

The woman is asked to lie in a ball on her side, and a hollow needle is inserted into her spine. Once in place, a catheter is threaded into the hollow needle, and the anaesthetic is injected into the catheter by means of a syringe. The drug lasts for two hours, but it can be topped up when necessary through the catheter.

The effect of an epidural is to numb the body from the waist down. Should the woman then need an episiotomy, forceps delivery, or Caesarian section, she will not need any further anaesthetic. The advantages are that the woman remains mentally alert during her labour. Epidurals are particularly valuable for women suffering from high blood pressure.

The disadvantages are that occasionally the epidural may not work at all, or it may take effect on only half the body. If the second stage of labour starts before the epidural has worn off, then the woman may find it difficult to push the baby out. She will need to rely on the midwife to tell her when to push, so there is a greater likelihood of a forceps delivery. Finally, there is the possibility of the woman developing a headache after the anaesthetic has worn off.

Gas and air This is a mixture of nitrous oxide and oxygen contained in a gas cylinder beside the bed in the delivery room. If the woman needs pain relief, she can put the rubber face-mask over her mouth and nose, and breathe deeply until the pain subsides. The gas takes effect after 20 seconds, and causes temporary drowsiness for about one minute, which is usually enough to help the woman through a contraction.

The advantages are that there are no harmful side-effects for the woman or her baby, but the pain relief may not be effective for everyone.

Other methods Other, more natural methods of pain relief are becoming increasingly popular. These include relaxation and breathing techniques, coupled with an understanding by the woman of exactly what is happening to her body during the birth. The belief here is that, if the fear is removed, there will be less pain.

Some women have found hypnosis to be valuable, but it is necessary to practise this during pregnancy to make sure that it is right for the individual.

Again, women who have been helped by acupuncture before pregnancy may find it helpful during labour.

Home delivery

Some women prefer to give birth at home and, assuming the pregnancy has progressed normally, most areas are able to cater for a home delivery. The advantages of giving birth at home are obvious – the surroundings are familiar and friendly, so the woman will be more relaxed and contented. However, most people – both members of the medical profession and parents – feel that the disadvantages outweigh the advantages. If there were any complications, the hospital has all the staff and equipment necessary to cope with an emergency. If an emergency arose at home, the woman would have to wait for an ambulance, which would waste valuable time.

In the case of a home delivery, the midwife is contacted as soon as labour begins. She will examine the woman and contact her GP, who may decide to come at this stage or wait until the second stage. The midwife will usually bring a portable foetal monitor and a gas-and-air cylinder for pain relief.

The second stage – the birth

Once the contractions have dilated the cervix to 10 cm, the mother will go into the second stage of labour – the actual birth, or delivery, of the baby.

In hospitals where there is a separate delivery room, the woman will be taken to it if she is not already there.

The second stage of labour is shorter than the first, lasting around 30 to 60 minutes for a first baby, and 15 to 45 minutes for a second and subsequent baby. The second stage is the point in labour when the woman feels she can at last do something positive, by pushing her baby down through the birth canal and out into the world.

In hospitals, most women give birth lying on their backs, as this has been the traditional birth position for many years. Many doctors, midwives, and mothers now believe that it may be easier for the woman if she is partly upright or squatting when she gives birth, as she will then have gravity on her side.

The urge to push, or bear down, is a reflex – at each contraction there will be about three to five urges to bear down. Although these pushes are relatively painless, they are very tiring, as they need to be controlled and gradual in order to reduce the risk of damaging the vaginal tissue. After each set of pushes, the woman should take some deep breaths to help her relax.

In a normal delivery, the back of the baby's head enters the birth canal first (fig. 4.3). During the second stage of labour, as the baby manoeuvres its way through the birth canal, it will do a 90° turn, so that its face lies in the curve of the mother's spine.

The 'crowning', which is the first sight of the baby's head, shows that delivery is near. The head may disappear again during contractions, but it is usually delivered within the couple of contractions following the crowning. At this stage, the midwife or obstetrician may decide to carry out an *episiotomy* (fig. 4.4) in order to help the baby out more quickly without damaging the mother. Occasionally, the tissue between the vagina and anus – the *perineum* – may tear as the baby's head is delivered. To avoid this, a small cut is made, as doctors believe that it is easier to repair a cut than a tear.

Many people, including some members of the medical profession, feel that episiotomies are carried out too frequently. The stitches can be very painful afterwards, and may cause discomfort for many months to come. Many doctors now carry out an episiotomy only if it is really necessary.

As the baby's head is born, the shoulders have to rotate in order to pass through the mother's pelvis;

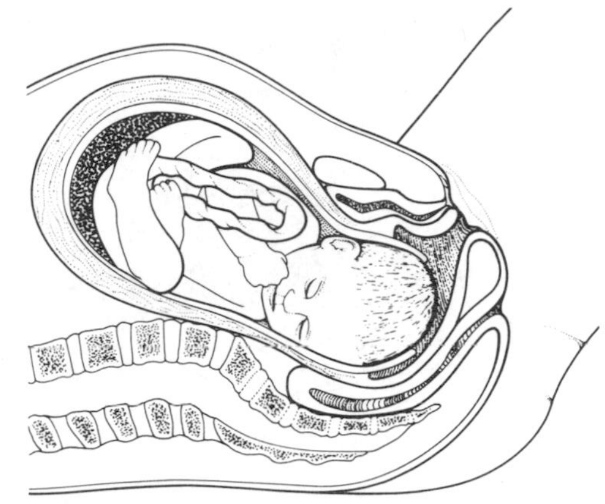

1. The baby is curled up in the foetal position while the cervix gradually dilates.

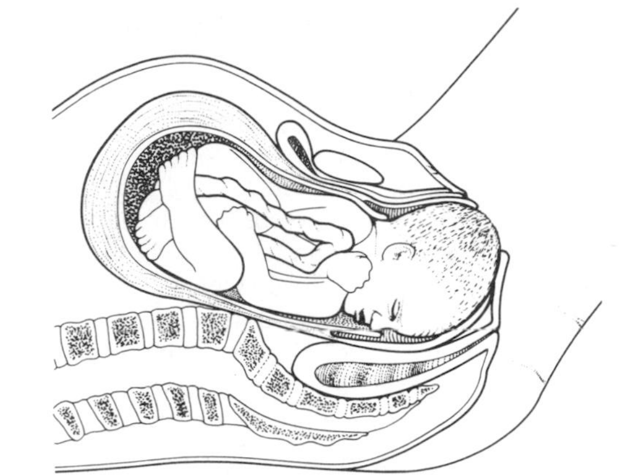

2. With the cervix fully dilated, regular contractions push the baby out along the birth canal.

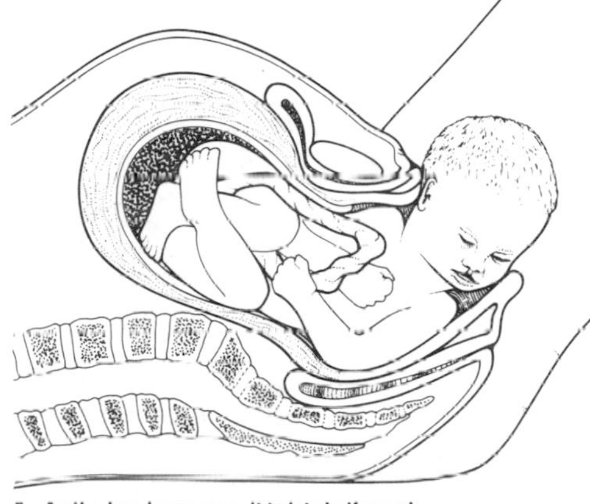

3. As the head emerges, it twists half round.

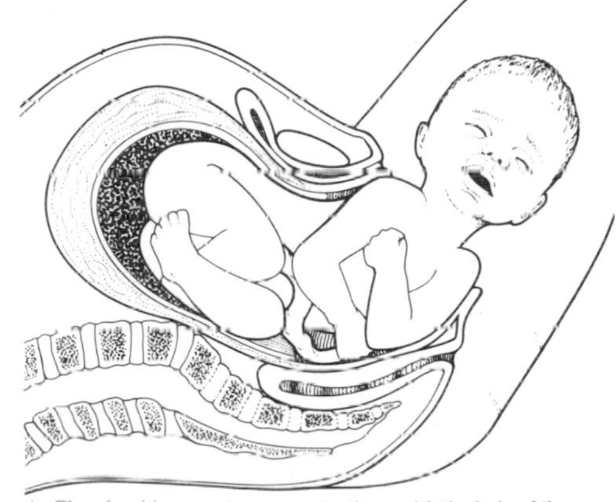

4. The shoulders are born one at a time, with the help of the midwife.

Fig. 4.3 The birth

then they are born, one at a time, quickly followed by the rest of the body. The placenta is still inside the mother, and attached to the baby by the *umbilical cord.*

Immediately after the birth, the midwife will clear any mucus from the baby's mouth and nasal passages to help it to breathe. The baby's umbilical cord is then clamped and cut and, very often, the baby is given to the mother to put to her breast.

The third stage

Around the time the baby's shoulders are born, the mother is given an injection in her upper thigh. The drug, usually syntocin, causes the uterus to contract and expel the *placenta*, or afterbirth.

Another reason for having the injection is to reduce the chances of the mother haemorrhaging

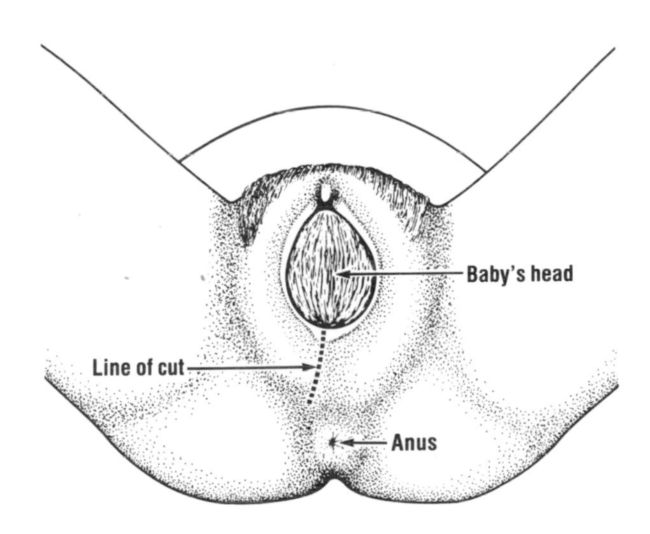

Fig. 4.4 Episiotomy

from the blood vessels that supplied the placenta with blood during pregnancy.

It is a good idea to put the baby to the breast as soon as possible after delivery, as the sucking stimulates the mother's uterus to contract, so helping the delivery of the placenta.

Until recently, it has been the custom to clamp and cut the umbilical cord soon after the baby is born. Nowadays, due to the influence of natural childbirth methods, it is quite common to wait until the baby has started breathing. The cord is clamped about 10 cm from the baby and again about 5 cm further on. The cord is cut between the clamps, and a rubber band or clip is clamped on the stump, which is then covered with a sterile dressing.

The baby is then checked to make sure it is healthy. If everything is normal, it will be cleaned up and an identity bracelet put on its wrist. Meanwhile, within 3 to 15 minutes of giving birth, the mother's uterus contracts to expel the placenta. The midwife helps by gently pulling on the cord with each contraction. Once delivered, the placenta has to be checked thoroughly to make sure it is complete. If any of it is left inside the uterus, it could cause an infection.

Once the birth is over, the mother, too, is checked to make sure she is well. If she needs stitches from a tear or episiotomy, she will be given a local anaesthetic and the stitches inserted. After her temperature, pulse and blood pressure have been checked, she is cleaned up, washed, and taken to the post-natal ward. Depending on the time of day and the procedure at the hospital, her baby will either be left with her or taken to the nursery.

With a straightforward home delivery, the procedure is the same. When the midwife leaves, she takes the placenta with her for disposal. She will come and check mother and baby for the next 10 days.

Complications during labour

Although most births are quite normal, there can sometimes be problems. Not all babies are born head first, and this is dealt with in the assignment section.

There may be other, unexpected problems that become apparent during labour, although the pregnancy may have been perfectly normal.

Forceps and vacuum extraction

Occasionally, it is necessary to use forceps (fig. 4.5)

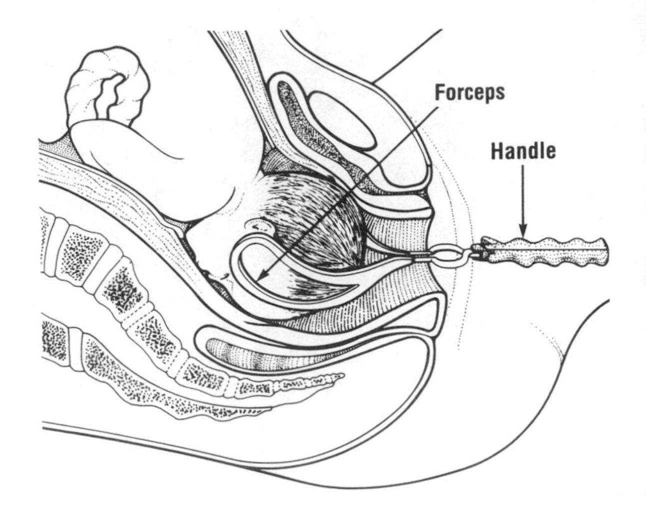

Fig. 4.5 Forceps delivery

to help the baby through the birth canal. Sometimes forceps may be needed because the mother has had an epidural, or other pain-killing drugs, that may prevent her from pushing efficiently. Before forceps are used, the mother will be given a pain-killing injection.

Occasionally, vacuum extraction (fig. 4.6), or ventouse extraction as it is sometimes called, may be used instead of forceps, especially if the woman has had an epidural. This method is used if the birth is expected to be straightforward, and the mother

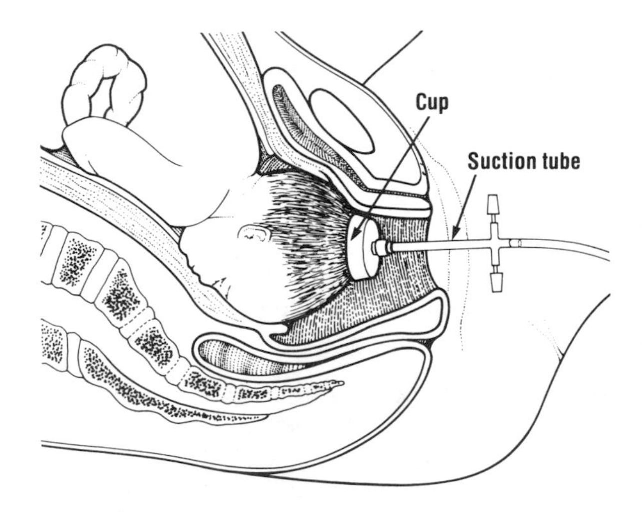

Fig. 4.6 Vacuum extraction

can help by bearing down. A small metal cup is attached to the baby's head by suction, and the baby is gradually pulled out.

In general, forceps and vacuum extraction are used to deliver a premature baby in order to protect its delicate head, or if there are signs of foetal or maternal distress.

A labour is considered to be prolonged if it lasts for longer than 12 hours, which applies to about one in ten of first deliveries. There are many causes for this, including the position of the baby, the shape of the mother's pelvis, inadequate contractions, and so on. It is sometimes possible to speed up a slow labour by giving a synthetic hormone, oxytocin, to encourage contractions. If this does not work, then it may be necessary to deliver the baby by Caesarian section.

Caesarian section

A Caesarian section is an operation by which the baby is delivered through a cut in the abdominal wall, rather than through the vagina. Sometimes it is known that a woman will need a Caesarian section before she goes into labour, and at other times it will be necessary due to unexpected complications during labour.

The operation is usually done under a general anaesthetic, so that the woman is completely unconscious. It is also possible to have the operation under local anaesthetic (an epidural), so that the woman is conscious, but not in pain.

Nowadays, the cut is made just under the 'bikini line' both for cosmetic reasons and because it heals quicker this way. The woman's pubic hair is shaved and, after the cut is made, the amniotic fluid is drained off. The baby is lifted out, and the mother's uterus and abdomen are stitched. The entire operation takes about 45 minutes, the last half hour being taken up with stitching. The stitches are taken out after about five days. Because the mother has had a Caesarian section, it does not necessarily follow that she cannot have her next baby vaginally, and the doctor will discuss this with her before she goes home.

Premature babies

If a baby is born before the 34th week of pregnancy, it may need to be put in an incubator to survive. As its lungs are immature, it may have breathing difficulties and, because it has little body fat, it may not be able to keep warm.

An incubator has a thermostat to keep the temperature even and the humidity constant. The baby's heartbeat, temperature, and respiration are monitored, and an alarm sounds if all is not well. The baby will be fed through either a vein or a tube leading to the stomach.

Overdue babies

Most babies are not born in the week they are due. This does not matter unless the pregnancy goes into 42 weeks. If at this stage the doctor thinks that the expected date of delivery was correct and the baby is overdue, he or she may decide to induce labour.

This can be done by rupturing the amniotic sac, when labour may follow naturally. Alternatively, a pessary containing hormones to produce uterine contractions may be put into the vagina.

The third method is to inject the woman with oxytocin. This method of induction is used if the mother's or baby's health is in danger. The mother may have high blood pressure or oedema, or the baby may be getting too little nourishment from an ageing placenta.

The baby after delivery

Babies are all different, and this is evident as soon as they are born. Some look very strange after birth, but most look fairly normal a few days later. The baby may be a bluish colour at birth – which is quite normal as it has not yet taken a breath. It may be covered in the whitish coating which protected it from the fluid inside the uterus, and its head may be temporarily elongated due to the birth process.

The baby may have one or more of these characteristics:

- There may be a swelling on the head, called a *caput*, which is caused by pressure as the baby presses on the cervix before birth. The swelling is not dangerous and disappears within a few days.
- The baby may be covered with *vernix*, which is the waxy, waterproof coating covering the baby's skin in the uterus.
- Many babies are born with dark hair over most of their bodies, particularly if they are premature. This hair, which is called *lanugo*, drops out in the first few weeks. The hair on the baby's head often drops out, and the new hair which grows may be a different colour.

- Many babies are born with a birthmark. The most common one is a pinkish mark over the eyelids – often called a 'stork's beak mark' – which usually disappears within a few months, but others may have to be treated later in life.

  Negro and Asian babies are sometimes born with a bluish mark on their backs. This is called a 'Mongolian blue spot' and usually disappears in the first year.

The mother after delivery

For the first week after delivery, the mother needs to rest as much as possible. The length of her stay in hospital depends on many factors, and the mother should be advised by the medical staff. If all is well with mother and baby, she can go home within 48 hours if she wishes.

After the birth, the uterus continues to contract, which may be slightly painful at times, particularly if the mother is breast-feeding. This is a good sign, as it means that the uterus is returning to normal quickly, and breast-feeding helps this process.

The uterus will continue to discharge blood and mucus for anything up to six weeks. This discharge, called the *lochia*, is quite heavy for the first few days, but gradually reduces. The bleeding stops far sooner if the mother is breast-feeding.

If she has had stitches during labour, she will find sitting up very uncomfortable. It may help her to sit on an inflated rubber ring, which takes the weight off the tender parts. It also helps if she has salt baths, as these speed up the healing process as well as helping to reduce the risk of infection.

The mother will need to pass water more frequently for a few days after the birth, as this is the body's way of removing the excess fluids that have built up during the pregnancy.

It is quite common for the mother to be slightly constipated after the birth, especially if she has had stitches. Unless the constipation is particularly bad, it would be better for her to eat more fibre, rather than taking laxatives. However, this may not be easy on a hospital diet.

Most women are disappointed by the shape of their bodies after the birth, as many expect to go back to their pre-pregnancy shape straight away. The stomach will be soft and sagging as there is no longer a baby inside to fill it out. The breasts will feel sore and swollen, and breast-feeding may cause cracked nipples. It is important to start on post-natal exercises as soon as possible, to help tone up the sagging muscles.

Many women find themselves left with purple stretch marks on their abdomens, but in time these fade to a thin silvery colour that is barely visible.

Because of the sudden change in the hormone levels in her body, the mother may suffer from the *baby blues* some time within the first five days after birth. Occasionally, the mother may develop a more serious form of depression, called *post-natal depression*. This can occur any time up to one year after her baby's birth. She will need medical help for this condition.

Assignments

1. Find out about the last few weeks of pregnancy by talking to mothers who have recently had babies. Ask them about their physical and emotional feelings.

   Perhaps it would be possible to ask someone in the late stages of pregnancy to come and talk to the group and answer any questions.
2. In pairs, decide what a couple need to buy in preparation for the birth of their baby. Bear in mind the following:

   - safety
   - economy
   - convenience

   There are many textbooks on pregnancy and birth to help you. The Health Education Council produces a booklet called *The Pregnancy Book*, which has a helpful section on what you need for the baby.
3. The parents need to decide where the baby will sleep. This may be a separate room, or part of another room shared with a brother, sister, or even the parents. Although the baby may not use the room for the first few weeks, it is necessary to plan it beforehand, as time is limited after the baby is born.

   In pairs, design a nursery suitable for a baby boy or girl. Include:

   - furniture and storage
   - essential baby equipment
   - decoration, furniture, and floor covering
   - toys, pictures, mobiles, etc. to interest the baby

   As before, bear in mind safety, economy, and convenience.

4. Find out what the expectant mother needs to get ready for the birth of her baby if she is:

   - having her baby at home
   - having her baby in hospital

   The information can be found either in books and leaflets, or by contacting a local midwife.
5. Check that you understand the meaning of these terms: electronic monitoring, foetal distress, enema.
6. A slow labour may lead to foetal distress. Find out the symptoms of foetal distress and what the medical staff may do to help the baby.
7. Research into the more traditional forms of pain relief available. Fill in a chart, using these headings:

Type of drug	Effect	Advantages	Disadvantages

   As well as finding out factual information, it would be valuable to talk to some mothers to find out their opinions. Mothers could be contacted at the local parent-and-toddler groups, playgroups, and through neighbours and the family. Decide beforehand on the questions you want to ask.
8. a) Find out the alternative methods of pain relief during labour. These addresses may be helpful:

   - Active Birth Movement, 32 Willow Road, London NW3 1TL
   - British Acupuncture Association, 34 Alderney Street, London SWIV 4EU
   - British Hypnotherapy Association, 67 Upper Berkeley Street, London W1H 7DH
   - National Childbirth Trust, 9 Queensborough Terrace, London W2 3TB
   - The Birth Centre, 101 Tufnell Park Road, London N7 0PS

   If you write for information, do enclose a stamped addressed envelope. When you have collected the information, discuss the advantages and disadvantages of natural pain relief versus the more traditional methods.

   b) Ask some mothers who have given birth without drugs to discuss their opinions.

   - Ask them when and why they made the decision.
   - Did they want to change their minds during labour.
   - Did the hospital encourage their decision?
9. There is a growing interest in natural childbirth methods, where the mother is allowed to give birth with as little medical intervention as possible. In France, Frédéric Leboyer did much to make natural childbirth more acceptable in hospitals in the seventies, and Michel Odent has developed his own ideas from this. As a result of this work, many of our hospitals are adapting to meet the needs of women who favour an active and natural birth.

   a) Find out about the work carried out by Leboyer and Odent.
   b) Is your local maternity unit able to offer active birth facilities? If so, find out what it offers. If not, find the nearest maternity unit which does offer this service.
10. Between the first and second stages of labour, some women experience what is called a *transition* period. In pairs, find out:

   - what the transition period is
   - what the woman can do to help relieve the discomfort
11. Although we have covered normal delivery in the text, not all births are completely straightforward. Here are three other positions, or *presentations*, in which the baby may be delivered:

   - breech
   - frank breech
   - transverse lie

   Draw a diagram to show each of these deliveries, and explain the difficulties each may cause.
12. Sometimes a mother may need a Caesarian section to deliver her baby. It may be necessary for any of the following reasons:

   - foetal distress
   - if the baby needs to be delivered early

- if the baby is too large for the mother's pelvis
- a vaginal infection, such as genital herpes
- a detached placenta
- placenta praevia

Find out the causes and symptoms of each of these (except foetal distress).

13. Many people are concerned that induced pregnancies are carried out for the convenience of the medical staff, in what is known as a 'nine to five' delivery. Discuss this approach to birth. Is induction a good thing, or should nature be allowed to take its course?
14. After delivery, the baby is given a check-up so that, if anything is wrong, it can be dealt with as soon as possible. Listed below are the tests carried out on the baby. Find out about each of them:

    - The Apgar scale, which tests the baby's heart rate, breathing, movements, skin colour, reflex response. This test is carried out after birth, and again five minutes later.
      Find out what the medical staff are looking for, and how the test is scored. Why is the test carried out twice?
    - The baby's legs are moved. What condition is being checked?
    - The genitals are checked for abnormalities. What minor problems might there be?
    - The roof of the mouth is checked. What abnormality could there be, and what problem would this cause a new-born baby?
    - The heart is listened to. What defect could be found?
    - The baby is weighed. What problems could there be with low birth-weight babies?

15. Although most babies are born healthy, some are born with special needs.

    a) Find out about special needs. It may be possible to invite a member of staff from the maternity unit to talk to the group about the subject. If this is not possible, then there is a great deal of information in books.
       Here are some examples of special needs – the baby may have one or more of them:

       - low birth-weight, either because the baby is 'small for dates', or premature
       - neonatal jaundice
       - breathing difficulties
       - low blood-sugar level (hypoglycaemia)

    b) Sometimes the baby will need special care in an incubator, where its temperature, pulse, heart-rate, and breathing can be carefully monitored. The baby may be fed with its mother's breast-milk, and the parents are encouraged to involve themselves with their baby so that they can form a relationship.
       Find out what happens in your local maternity unit:

       - Is there a special-care baby unit?
       - What equipment is available for special needs?
       - How much are parents involved?

16. Many mothers suffer from the 'baby blues' after their baby is born. A minority suffer from post-natal depression. Carry out some research into these two conditions. Find out:

    - the symptoms
    - the possible causes
    - what help is available

    Perhaps it would be possible to talk to some mothers about their experiences.
17. Put together a questionnaire to ask your local maternity unit about the services it offers. Write your own questions, but include these points:

    - What is the admissions procedure (i.e. what questions is the woman asked, what tests are carried out, etc.)?
    - Is the father's participation actively encouraged?
    - Is there an active birth option?
    - Is foetal monitoring available and, if so, is it routine?
    - When the woman is in labour, does she go into a first-stage room or into the delivery room?
    - In general, how long do mothers stay in hospital after delivery?
    - Are epidurals available?
    - Are episiotomies given as routine, or only when necessary?
    - Assuming the health of mother and baby

are good, how long over the estimated date of delivery is a woman allowed to go?

18. Post-natal exercises are essential, not just to improve the mother's figure but also to improve the tone of the muscles in the pelvic floor. As a group, obtain a sheet showing post-natal exercises, and try them out. Feel which muscles each exercise works on.

5 Caring for the new baby

Finally, after months of waiting, the baby is born. This is an emotional time for most couples, as they are now responsible for a new life. The baby, too, has some adapting to do. For the first time in her life, she has to live outside the mother and breathe, feed, digest, and communicate for herself.

Ante-natal classes should have shown the new mother how to cope with the everyday care of her baby, how to change nappies, feed, and keep the baby clean. Booklets will tell her how to recognise signs of illness, and friends and family will offer plenty of advice.

The first six weeks are a time of learning by trial and error, and adapting. The baby is new, and no one knows how she will react to situations. The mother, having to cope with changes in her own body, will be tired as well as concerned about whether she is doing the right thing for her baby. Ideally, the mother and her baby will develop a special relationship, known as *bonding*.

Fortunately, the importance of encouraging bonding as early as possible is recognised, and in hospitals efforts are made to help. After the birth, the mother may wish to hold her baby close and put her to the breast. Breast-feeding will also help her body return to normal as quickly as possible.

Bonding is also important for the father, and he too needs an opportunity to hold his new-born child close, especially as in most homes the father tends to have less contact with the children than the mother.

The new-born baby has certain reflexes, which are covered in more detail in the assignment section. One of these reflexes – the rooting reflex – reflects the baby's survival instinct to feed.

Crying

The new baby communicates all her needs by crying. Babies always cry for a reason, even if it is only because they want a cuddle, and the mother is doing the right thing when she listens to her baby's cries.

There are quite a few reasons why babies cry, and it takes time for the parents to distinguish between the various cries, but it is up to them to find out what is wrong.

Obviously the main cause of crying is hunger, and to satisfy this, the baby needs milk.

A baby will get very upset if, for instance, her nappy is changed when she is hungry, so the timing of baths, nappy changes, and so on should be thought about carefully. Bath-time should be a pleasure for parent and baby, so a time when the baby is awake, but not hungry, should be chosen.

Most parents can tell when their baby is crying from shock or fear, as the cry is urgent and wavering. This type of crying is usually caused by a shock such as a loud noise, a bright light, or over-excitement.

Many babies hate being dressed and undressed, because they feel very vulnerable. Parents can help by keeping close physical contact, which will reassure the baby.

Sometimes, a baby simply needs company and a cuddle and will cry as soon as she is put down. Nowadays, there are still people who believe it is wrong to 'give in' to the baby's demands, but most parents are not happy to leave their baby to cry. One way round this problem is to keep the baby in a baby-sling, where she can enjoy closeness and security.

Some babies seem to need a lot of sucking, and they may be satisfied by a dummy. Most mothers who use a dummy buy it as a last resort: they hate the look of the dummy, and feel they have failed in some way. However, used properly, dummies do the baby no harm and satisfy the need to suck. The disadvantage of dummies is that it is easy to over-use them – every time the baby cries, the dummy is put in her mouth. If the baby only needs a dummy to help her go to sleep, then that is when the dummy should be used. If the baby has the dummy too often, she will become dependent on it, and it is quite difficult to wean a two- or three-year-old off a dummy.

Sometimes a baby will cry in the late afternoon or evening, and the crying will be very intense. Often the baby will draw her legs up at the same time. This is thought to be caused by *evening colic*, or *three-month colic*. The condition usually starts when the baby is about three weeks old and stops when she is about 12 weeks. With true evening colic, it is impossible to pacify the baby. The cause of this colic is not really understood, but it could have something to do with the baby's having an immature digestive system. There is no treatment, and the parents can only look forward to the fact that it will pass, and try temporarily to plan their everyday lives to cope with these troublesome few hours.

Sleep

Many parents worry about the amount of sleep their baby should have, and how to establish a sleeping routine. If the baby is happy and thriving, then she is getting enough sleep.

In the first few days after birth, the baby will doze on and off for much of the time. Once back at home, it is a good idea to develop a routine as soon as possible, so that the baby can distinguish between being awake and being asleep. This means the baby will be more likely to sleep longer at night.

If the baby seems sleepy, then she should be put in the pram or cot. If the baby is awake, she should be got up. In time, the baby will associate places to sleep in and places to be awake in. There is no need to creep around the house for fear of waking a sleeping baby – if the baby is used to complete silence as she sleeps, then in time she will be unable to sleep if there is any noise at all.

It is a good idea to make a bed-time routine, so that the baby associates it with a long sleep. If a baby is fed at night, the night feed should be different from the day feed, with no playing or talking to stimulate the baby into being wakeful. The parents should try to keep the lights low, and feed the baby with as little fuss as possible.

It is not wise for the baby to sleep by the parents' bedside all night, as the parents will find that they cannot sleep very well and will wake up every time the baby snuffles or moves.

The senses

In the early days, the baby learns by perception – that is, through the senses of sight, hearing, taste, smell, and touch. While the baby is awake, the parents can help her to develop these senses.

A new-born baby can see clearly for about 20–25 cm, but anything beyond that is blurred. In time, the baby will recognise the mother's face, and associate her with the pleasures of feeding and cuddling. Research has found that babies are drawn towards the human face more than anything else.

It is thought that babies can sense sounds while they are still in the womb, and some people believe that a recording of a heartbeat will soothe a crying baby. At birth, a baby can hear all the sounds around her: loud noises will make her jump and cry, whereas rhythmic sounds will be soothing.

Listening is not the same as hearing, as listening is a conscious skill, and it is listening that leads to learning. A baby will listen positively to the sound of a voice. She will associate the sound of her mother's voice with pleasure, and will react by becoming excited.

As the baby's skills develop, she will try to communicate by making noises, often after being spoken to.

The baby is born with the senses of taste and smell. A baby will react to unpleasant tastes, and in experiments will prefer to suck sweetened to unsweetened water. It is thought that she knows the smell of her own mother.

Breast-feeding

Most women have decided whether they want to breast- or bottle-feed before their baby is born. Nowadays breast-feeding (fig. 5.1) is encouraged, because of the advantages for the mother and the baby. It may take more time to establish breast-feeding, but most of the problems can be solved.

The breast-feeding mother will need a special supportive bra that fastens at the front. The type that completely exposes the breast is best, as then the baby can make close physical contact with her mother. For the first few weeks, the mother should wear the bra day and night to stop her breasts from dropping and sagging.

For the first two to three days of breast-feeding, the mother produces *colostrum*, a fluid which is high in protein, vitamins, and minerals, and

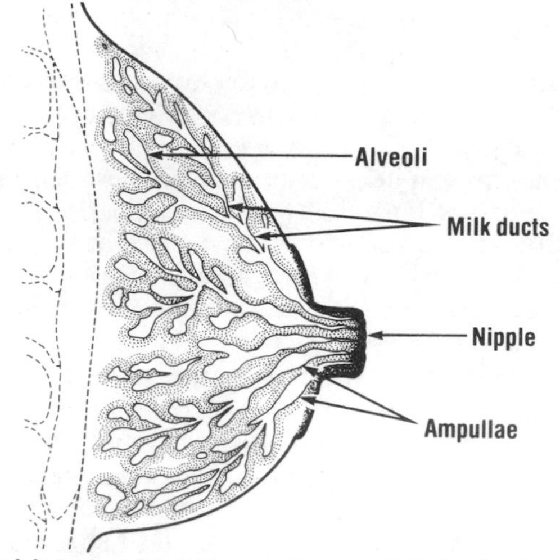

(a) Each breast contains many small sacs called alveoli, which produce the milk. The milk travels along the ducts and collects in reservoirs called ampullae. When the baby sucks, the milk comes out through the nipple.

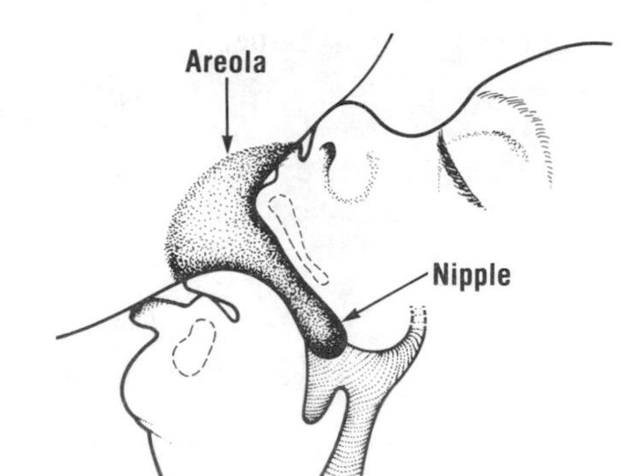

(b) For successful breast-feeding, the nipple should be well into the baby's mouth.

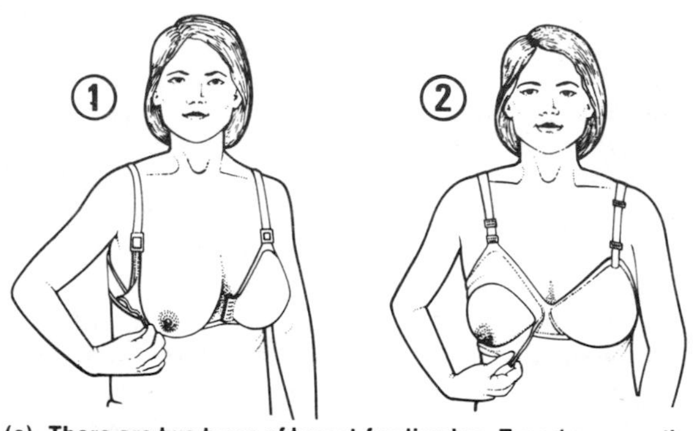

(c) There are two types of breast-feeding bra. Type 1 exposes the whole breast, which is nicer for the baby. Type 2 exposes only the nipple.

Fig. 5.1 Breast-feeding

contains many of the mother's antibodies. The fluid is very nourishing, and helps protect the baby against infection.

As the baby sucks, the breast is stimulated to produce milk. By about the third day, the milk comes in and the breasts may become sore and swollen. The best way to reduce the discomfort is to put the baby to the breast.

Leaking may be a minor problem to start with, but breast-pads can be put inside the bra to absorb any excess milk and prevent it from leaking through to the clothes.

After a few weeks, breast-feeding becomes established, as the amount of milk the mother produces adapts to the demands of the baby.

In the past, babies were fed four-hourly, and it was thought that feeding them more often was bad training. However, the best way to feed a baby is 'on demand.' After all, the baby knows best when she is hungry. Few hospitals now limit feeding times, and this makes for a more relaxed and successful start to breast-feeding. A mother who feels tense will find feeding her baby very difficult.

The hospital will recommend giving the baby

boiled water as well as milk. When breast-feeding, this is not necessarily important once the mother goes home, assuming the baby produces wet nappies. Hospitals tend to be hot, so the baby may not want the water once at home.

Breast-feeding is a case of 'supply and demand'. The more the baby sucks, the more milk is produced. It follows that a mother who gives her baby a bottle-feed after a breast-feed to 'top her up', will find that her own milk supply diminishes. This then means that the baby will need more bottle top-ups, and breast-milk production will gradually diminish. The same applies to mothers who start solid foods too early (before four to six months). The baby gets nourishment from the food rather than the milk, which again reduces the breast-milk supply.

Breast-feeding does have some difficulties, and the mother should be given plenty of support to overcome these. In rare cases, the baby's handicap may prevent successful breast-feeding. Far more common are the minor problems of engorgement, sore or cracked nipples, blocked ducts, and so on – these are covered in the assignment section.

Care of the breasts before and during breast-feeding can help reduce the problems that may arise, and a relaxed attitude will help milk-production. As time passes and the mother returns to her everyday household chores, she may find her milk diminishes. In order to increase her production of breast-milk, she could follow these guidelines:

- put her feet up as often as possible, especially at the end of the day
- have plenty to drink, to make up for the fluid she loses by breast-feeding
- make sure that the breasts are emptied after each feed, by expressing the milk if necessary
- make sure she eats a proper diet

Bottle-feeding

Bottle-feeding does have some advantages, as it means that the mother is not the only person who can feed the baby, and this gives her greater freedom. It also means that the father can become more involved in feeding his baby.

However, cows' milk is made for calves, not human babies, so it needs to be carefully processed to make it closer to human milk. In general, the more formula milk costs, the closer it is to human milk, and the better it is for the baby.

Apart from the cost of formula milk, special equipment (fig. 5.2) needs to be bought: the bottles and the sterilising unit (although it is possible to adapt other containers to make a sterilising unit).

Hygiene is very important when bottle-feeding, especially during the first few months, when infections can be very serious. A new baby has little immunity, especially if she has never been breast-fed. Milk left at room temperature is an ideal breeding ground for harmful bacteria, which can cause gastroenteritis.

Before making up a bottle, you should wash your hands, and everything used to make up the feed – bottles, teats, teat-covers, jug, spoon, and water – should be sterile. The equipment can be sterilised by boiling it for 10 minutes or, more simply, by soaking it in specially prepared sterilising fluid.

The instructions for making up the milk should be followed carefully. Research has shown that many feeds are not made up properly, and this can be bad for the baby. The powdered milk should be measured accurately and the excess powder levelled off in the scoop with a knife.

Liquid milk concentrate should be measured at eye level to be sure of an accurate reading.

Water should be measured after it has boiled, not before, as some will evaporate off as steam.

If there is a fridge available, a day's feeds can be made up together, and these can be taken out and heated up as necessary. It is important to keep the made-up milk ice cold: warm milk should never be carried around for any length of time, as bacteria can breed. Each bottle should be heated as it is needed.

Although it is impossible to overfeed a breast-fed baby, it is quite common for bottle-fed babies to gain weight too quickly. Research has shown that fat babies tend to become fat adults, so bottle-fed babies should not be overfed. In the past, fat babies were considered bonny and beautiful, but today opinions have changed.

To avoid overfeeding, the feed should be made up in accordance with the manufacturer's instructions.

Sometimes a baby's cries of thirst are mistaken for hunger, and another feed is given instead of a drink of boiled, unsweetened water. This will soon lead to excess weight gain.

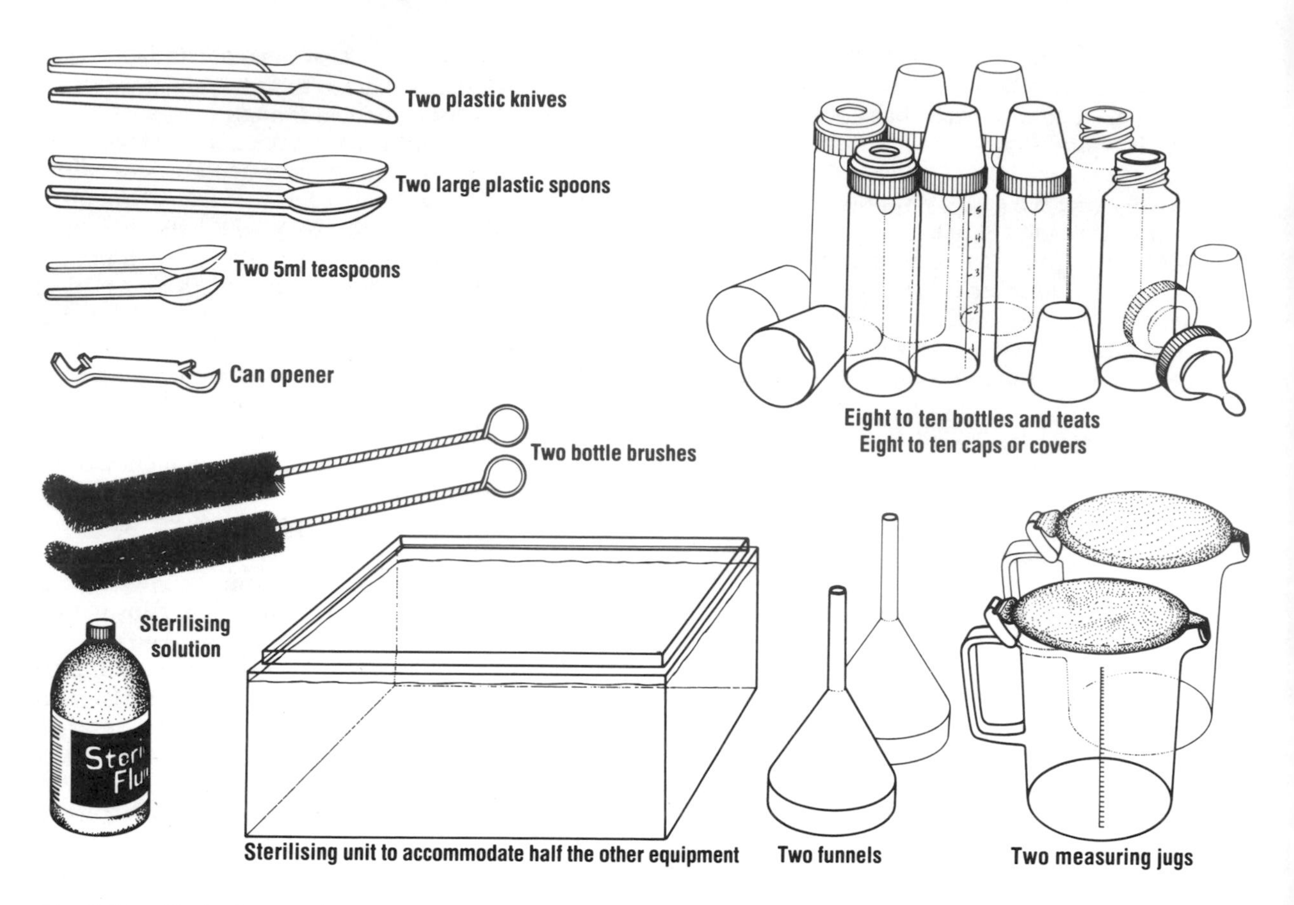

Fig. 5.2 Bottle-feeding equipment

Wind

When a baby cries, people often say it is wind. The danger is that mothers become unnecessarily worried about wind, and will spend wasted time patting and rubbing the baby's back, waiting for the wind to come up. All babies swallow some air with their feed, and sometimes the air may cause some discomfort. In this case, the baby should be given a chance to burp by being held over the shoulder for a few moments.

Sometimes, a mother will lay the baby across her knees to burp, but this is likely to make the baby bring up some milk as well. Another poor way of burping a baby is to sit the baby on the lap. This folds the baby's stomach over and makes it very difficult for the baby to burp.

There is no danger in putting a baby that hasn't burped in her cot, although it is wise to lay the baby on her side or front, so that any milk that is brought up can drain out of the mouth easily.

Possetting

This is when the baby brings up milk during, or just after, feeds. Some babies are more prone to possetting than others, but there is usually nothing to worry about. The most common cause is overfeeding, so the baby brings up the excess milk. Sometimes the baby will take in too much air with the feed, so milk is brought up as the air escapes. The amount of milk brought up looks more than it actually is, but it is a good idea to have a cloth nearby to mop it up. A bottle-fed baby will become quite smelly if the possetted milk is not cleaned up.

Although many babies regularly bring up milk after a feed, the mother should be aware that it can be the start of a stomach complaint. Occasionally, a baby may vomit with great force – the milk may travel a metre or more. If this happens often, it can be a sign of a condition called *pyloric stenosis*. This is caused by a malformation of the muscles around the bottom of the stomach, and occurs more often in boys than girls.

Clothes and nappies

Many parents find that their babies hardly wear any of the first-size clothes because they grow out of them so quickly. They may also find that the pretty baby clothes, with lacy patterns and ribbons, are a nuisance as they are likely to trap the baby's fingers.

The mother needs to decide whether to choose terry or disposable nappies. Although disposable nappies are expensive, they are very convenient and hygienic. To work out the cost, the mother should reckon on about six nappy changes a day.

The best time to change a nappy is after a feed, when the baby is more likely to have dirtied the nappy. Everything should be at hand before changing the nappy (fig. 5.3). A changing mat is useful, as it leaves both hands free to deal with the baby. Other equipment needed is cotton wool, baby lotion, nappy-liners, pins, and creams.

A general nappy-changing routine consists of filling a nappy-bucket with sterilising fluid, to prevent any bacteria from soiled nappies causing nappy rash. Dirty nappies are soaked in this solution for at least six hours, and then rinsed. Many mothers find that this is not enough to keep their nappies looking clean (although they are clean), and prefer to wash them in the machine as well. As the sterilising fluid is poisonous, the mother must remember to wash her hands after dealing with the nappies.

Once the dirty nappy has been taken off the baby, her bottom should be cleansed with either warm water or baby lotion, with the mother wiping away from the genitals to prevent infection. After this, any creams are applied before the clean nappy is put on. Nappy-liners are useful because they collect the motion, which can then be flushed down the toilet.

In view of this tedious procedure, it is easy to see why more and more mothers are choosing the convenience of disposable nappies.

Excretion

Many mothers worry about their baby's bowel movements, or *stools*. Because the baby is new, her digestive system is adapting to new feeding patterns, so the stools vary in both colour and consistency. After about three weeks, the stools settle down. Breast-fed babies will tend to have soft, odourless stools, and may pass them at every feed. Bottle-fed babies have more solid stools, which usually smell more and are passed less frequently. Unlikely breast-fed babies, bottle-fed babies can become constipated. This is usually relieved by giving water to drink.

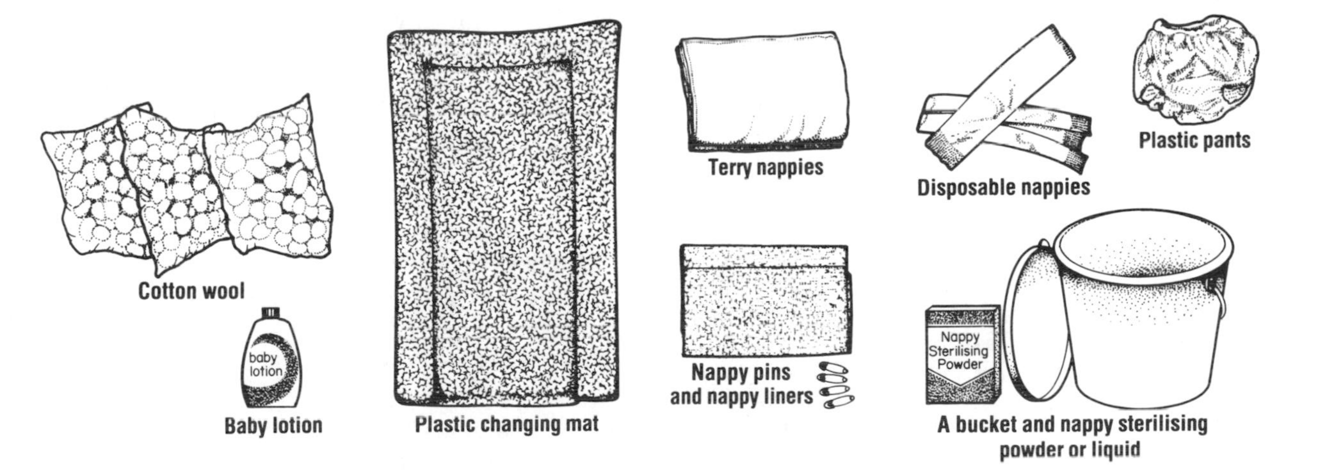

Fig. 5.3 Nappy-changing equipment

It does not matter how wet a baby's nappy is, as long as it is not dry. If the nappy is completely dry, the baby should be given extra fluid. If this does not work, the doctor should be contacted, as there may be a blockage somewhere. If the urine seems strong, then the baby should be given more to drink. This may happen in hot weather, or if the baby is developing a fever. Again, if a drink does not cure the problem, the doctor should be contacted in case there is an infection.

Nappy rash

Nappy rash is caused by ammonia in the urine acting on the baby's delicate skin. There is no fail-safe way of preventing nappy rash, but the likelihood can be reduced with special care.

The nappies should not be washed in biological powders. They should be thoroughly rinsed, dried, and aired before using, as dampness is ideal for encouraging nappy rash.

The baby should be left without a nappy on whenever possible, as this allows the air to get to the affected area. It also gives the baby an opportunity to kick freely without a bulky nappy on.

Barrier creams, such as zinc and castor oil, are soothing and help create a barrier between the urine and the baby's skin. If the baby has a bad nappy rash, she should be taken to the doctor's.

Keeping the baby clean

The baby's face, hands, and bottom should be cleaned twice a day – this is called *topping and tailing*. Although the baby does not need to be bathed every day, it can be a pleasure for parents and baby, and it can be part of the bed-time routine.

The baby should be bathed in a warm, draught-free room, and everything that is needed should be got ready beforehand. It is unwise to bath a baby at a time when she is hungry or miserable, as this will not encourage her to enjoy bath-time.

Many babies develop *cradle cap*, which is when dry scaly skin collects on the scalp. This often happens when parents are unwilling to wash the fontanelle properly, as they are frightened of damaging the baby. The fontanelle is the soft spot on top of a young baby's head where the skull bones have not yet joined together. If the baby's hair is washed about three times a week, and brushed every day, the stimulation will help prevent cradle cap.

The ears should be kept clean with a damp cotton bud, but inside the ear should never be touched, as there is a risk of damaging the ear drum. The hands should be cleaned with warm water, as dust collects in the folds of skin. The nails should be regularly cut to prevent the baby from scratching herself.

Warmth

New babies need to be kept warm, as their bodies are not able to keep warm as adults do. Their system of temperature control is immature and unable to cope with extremes of heat or cold. It would be too expensive to heat the entire house for a new baby, but for the first few weeks, the baby's bedroom should be kept well heated. It is better to keep a baby warm with layers of light clothing, or covers, rather than with thick, heavy ones, as the warmth will collect inside the layers.

Heat, too, can be damaging. The baby's delicate skin can easily be hurt by sun, wind, and direct heat from fires, and so on. Adults need to check for these dangers.

Minor illnesses

The first few weeks of a baby's life are a worrying time for parents, who may be anxious that they might miss any signs of illness in their baby. Generally, it is obvious when a baby is ill – she will be miserable and restless, and may be unwilling to feed. The baby should be taken to the doctor's unless she is very ill with a high temperature, in which case the doctor should be contacted immediately, as there is a risk of convulsions.

If the baby has sickness and/or diarrhoea, she should be given plenty of fluids to prevent dehydration. If the condition is bad, or lasts more than 24 hours, the doctor should be contacted.

Assignments on the new baby

1. The baby is born with certain reflexes. Find out about these reflexes, what they are, and when they are likely to disappear:
   - rooting reflex
   - 'walking' when held upright
   - hanging limply when held horizontally
   - legs curled up under body when put face down

- Moro reflex
- gripping with fingers

2. It is important to establish a sleeping routine, as a new-born baby has no concept of night or day. Work out a suitable bed-time routine for a new-born baby.
3. Babies first learn through their senses. Find out about the five senses:

   - Bearing in mind that a baby focuses on near objects best, think of some interesting toys, pictures, or objects that would help to stimulate the sense of sight and help the baby learn about her surroundings.
   - See if you can find any information or research on babies' sight.
   - As a group, see if there is anyone who has access to a new baby. If so, ask if the mother would be willing to bring the baby in for the group to observe his or her behaviour. If this is possible, make these observations:

     – How does the baby react to the mother's face?
     – Does the baby focus on the eyes more than any other part of the face? If the mother moves her face, does the baby follow with his or her eyes?
     – Does the baby seem interested in bright objects such as a window or light?

     Ask the mother to talk to her baby, and make these observations:

     – How does the baby react?
     – Does the baby react in the same way if someone else is talking?
     – Does the baby follow the sound of the mother's voice?

4. Discuss the advantages and disadvantages of breast-feeding and bottle-feeding.
5. Many mothers suffer from one or more of these breast-feeding problems:

   - engorgement
   - sore nipples
   - cracked nipples
   - abscesses
   - blocked ducts

   Explain the symptoms and causes of each of these, and find out how the discomfort can be treated.

6. Arrange to look at the wide variety of baby milks available. Before doing this, make a checklist of the questions you want to ask about each product:

   - make, price, and quantity in the packet
   - whether it is in dried, liquid, or ready-mixed form
   - whether sugar is added, or needs to be added
   - a comparison of added vitamins and iron
   - how it is made up
   - whether it is suitable for babies of all ages

7. Breast-feeding creates many responses from women. Discuss the following:

   - 'I couldn't breast-feed because my milk was too weak.'
   - 'I didn't have enough milk for my baby.'
   - 'I felt like a cow, so I changed to the bottle.'
   - 'The thought of breast-feeding is repulsive.'
   - 'I had to stop breast-feeding because my husband got jealous.'

8. A breast-feeding mother should take good care of herself in order to ensure that she produces enough milk for her baby. As a group, answer these questions:

   - Why should the breasts be emptied after each feed?
   - Why does the pill reduce the milk supply?

   Work out a day's diet for a breast-feeding mother.

9. Make a chart showing the essential equipment and clothing needed for a new baby. The following example may be helpful:

Item	Description	Approx. cost	Comments
4 vests	Stretch-towelling all-in-one with poppers for nappy change	£1.50	Good design as they don't ride up, and keep nappy in place. Comfortable.

   If you ask in advance, a local baby shop may be willing to let the group go on a visit and look at the baby equipment. Alternatively, many shops produce a catalogue of their goods.

10. Research has shown that many formula feeds are made up incorrectly. Feeds have been found to be too weak, too strong, or too sweet. What are the dangers of mixing a formula feed incorrectly?
11. Ask any mothers you may know to talk to you about wind in babies. Note down any interesting comments, and bring them along for the group to discuss. Were any of the comments clearly wrong? Were any of them valuable advice to a new mother?
12. Try to obtain some terry nappies and some life-size baby dolls, and practise the different methods of folding nappies.
13. Ask your local midwife or health visitor to demonstrate how to bath and 'top and tail' a baby. If you are unable to get hold of a real baby, then use a doll instead.
14. It is important to recognise when a baby is unwell. Find out about the following:
    - how to take a baby's temperature
    - how to reduce a temperature
    - what dehydration is
    - what the symptoms of jaundice are

After the birth – the parents

Pregnancy is a time of waiting and planning ahead. The couple expecting their first baby may have some idea of what to expect, as they may have read books, listened to friends and family, and attended ante-natal classes. However, once the baby has arrived, reality may prove very different from their expectations.

Feelings

The mother who is suffering from the 'baby blues' will need support from her partner, friends, and family. If the depression does not pass, she should talk to her doctor.

It is very easy to forget the feelings of the new father, as it is the mother and baby who tend to get all the attention. Fathers, too, feel emotional, but in our society, men are not encouraged to show their feelings. If his partner is in hospital with the baby, the man will be alone and separated from his new family, and would appreciate contact with friends and relatives.

It is important for the father to form a bond with his baby right from the start by taking an active part in child care. If there are older children in the family, he will probably have to look after them while his partner is in hospital, and this will strengthen his relationship with them.

It may take longer for a father to develop an attachment to the new baby, because he tends not to be so involved in the everyday care as much as the mother.

If there are older children in the family, their feelings need to be considered after a new baby is born. From now on, they are going to get less attention from their parents – especially from the mother, who will be spending a great deal of her time with the new baby. They need to feel that they are still loved and wanted, so they should be involved as much as possible in caring for the new baby.

Appearances

Many women put on weight during pregnancy and do not lose it after the baby is born. Post-natal exercises and healthy eating should help them to lose the excess weight within months.

Sex life

After having a baby, the couple need time to adjust to the changes in their lives. All parts of their life together are affected, including their sex life. The problem pages of women's magazines frequently print letters from readers who complain that they have never enjoyed sex since the birth of their baby.

This is a sad situation which can be avoided if a couple are willing to take time to find out each other's changing needs.

Many women lose their *libido*, or sex drive, after the baby is born. There may be one or more causes. A wakeful or demanding baby will leave little opportunity for the couple to make love. In addition to this, the woman will be very tired from coping with a difficult baby.

Other mothers are unable to change from the role of mother to that of lover, as they are so wrapped up in the baby.

Physically, the new mother may be uncomfortable as a result of an episiotomy, and the soreness will last for at least two to four weeks. If she is breast-feeding, her breasts will be sore and tender to start with, and she will not want any pressure on them. The couple should discuss these difficulties, and find positions that are comfortable.

If the woman finds intercourse painful, she should see her doctor in case there is an infection.

Family planning

As soon as the couple start their sex life again, it is possible for them to conceive another baby, even if the mother is breast-feeding. Although breast-feeding does reduce the chances of conceiving, it is not, as many people mistakenly believe, a reliable method of contraception. As a woman ovulates before her period, she could be pregnant and not realise it.

It is generally accepted that it is best to wait at least a year before conceiving another baby. This gives the woman's body a chance to return to normal, and it also gives the baby time to enjoy as much of the mother's attention as possible. To reduce the chances of an unplanned pregnancy, new parents need to think about which method of contraception suits them. (See chapter 20.)

Post-natal check-up

About four to seven weeks after her baby is born, the mother has a post-natal examination. This is done for three reasons:

1. to check that the mother's body is returning to normal after the birth
2. to check that the baby is making good progress
3. to give the mother the opportunity to discuss any problems or worries that she may have

The examination will include checking her blood pressure, breasts, and weight and making sure that the uterus has returned to its normal size. If she has had stitches, the scar will be examined to make sure it has healed. Some areas offer a cervical smear to check for cancerous cells. Any woman not offered a smear should ask for one, as cervical cancer is affecting younger women nowadays. The woman will be asked about contraception, and if she has decided to have the coil fitted, it will be done at this stage.

Assignments on the new baby: the family

1. How could you help an older child accept a new baby brother or sister into the family? Write some guidelines to help a couple who have just had their second child. Their first child is now two and a half years old.
2. Fatigue is one of the main problems a new mother suffers from. Write a checklist of thoughts and ideas that could help her to cope with tiredness. The ideas can be simple, such as energy-giving snacks, or more complex, such as getting a home help. Remember to check that all the ideas are feasible.
3. *Case study*

   'Sarah Matthews is a 20-year-old unmarried mother. When she told her boyfriend she was pregnant, he didn't want her to have the baby and told her to have an abortion. Sarah couldn't face this and decided to keep the baby.

   Her parents were sympathetic and gave Sarah plenty of emotional support. However, she decided she needed to be independent, and managed to get a flatlet in a council hostel for unmarried mothers. Sarah's main complaint is the lack of information available for single parents. It was only by trial and error, and a lot of hard work, that Sarah found out the benefits she was eligible for and what help was there for people in her situation.'

In pairs, compile a dossier on what is available, nationally and locally, for people in Sarah's position. Perhaps the following ideas may help:

- emotional help – where to go for emotional support
- economic help – the rights and benefits available
- local provision – day nurseries, crèches, housing help, etc.
- national provision – some addresses:

Gingerbread, 35 Wellington Street, London WC2E 7BN

Meet-a-mum Association (MAMA), 26a Cumnor Hill, Oxford OX2 9HA

National Council for Carers and their Elderly Dependants, 29 Chilworth Mews, London W2 3RG

National Council for One Parent Families, 255 Kentish Town Road, London NW5 2LX

Single-handed Ltd, Thorne House, Hankham Place, Stone Cross, Pevensey BN24 5ER

Families Need Fathers, B.M. Families, London WC1N 3XX

Scottish Council for Single Parents, 13 Gayfield Square, Edinburgh EH1 3NX

If you write to any of these addresses, be sure to enclose a stamped addressed envelope for a reply.

6 Child development in the first year

The first six months

After about six to eight weeks, life with a baby becomes more settled. The mother feels more like her old self again, although she may still be tired. The novelty of having a new baby will have lessened and, to some extent, a daily routine will have been established. In most cases, the mother will be able to predict the needs of the baby, and feeding patterns will be established. Breast-feeding mothers will find that feeding has become much easier, and there are fewer problems.

Solid foods

Introducing a baby to solid foods can be a difficult time, mainly because there are so many different views and opinions on when and how to begin mixed feeding. Solid foods should not be the start of *weaning*, because milk should still be the baby's main source of nourishment for many weeks to come.

As a general guideline, solid food should not be offered until the baby is at least 12 weeks old and/or weighs 5 kg. At this stage, solid food is not a necessity and should only be a taste to prepare the baby for her future diet.

As solid food is not replacing the milk in the baby's diet, the amount offered should be very small. All the vital nutrients are in the milk, so the solid food only gives extra calories for energy. Too much solid food will make the baby fat.

The first foods can be either cereals, puréed fruit, or commercially produced baby foods. As long as there are no lumps, the baby should be able to digest the food.

Timing is important when introducing solid foods, as the baby's natural instinct is to suck to satisfy her hunger. It is not a good idea to give solid food when the baby is very hungry – in the morning for instance – as the baby will cry with frustration at not being able to suck and will be unable to eat.

As the baby gets older – about five to six months – she will begin to associate spoon-feeding with the pleasure of satisfying hunger. In time, more nourishment will come from solid food and less from the breast or bottle, which means that the baby is weaning herself. It will be possible to see a pattern in the baby's feeding, for instance:

early morning: breast or bottle for comfort
breakfast: cereal and drink
lunch: savoury and sweet plus drink
tea: egg, bread and butter, and drink
bed-time: breast or bottle for comfort.

The baby will also be able to go for longer between feeds.

Organisation is important, as feeding-time can be very messy. The baby will want to touch the food and try to feed herself with finger foods such as rusks, bread, and fruit. This is all part of the process by which the baby learns to feed herself.

Hygiene is essential, as a stomach upset can be very dangerous for a baby of this age (see the assignment section).

Sleeping

By about six weeks, the baby's sleeping pattern will become more established. Some babies may sleep through most of the night quite early, whereas others may continue to have a night feed for some time.

The pattern of sleeping and only waking to feed will change as the baby stays awake for longer. This is a good chance for the parents to play with their baby, which will help her to develop and communicate.

There are some babies who need very little sleep. This is distressing for the parents, but does not seem to harm the baby.

Physical development

Physically, the baby is developing very quickly, as the table (fig. 6.1) for the first six months shows.

Fig. 6.1 Development up to six months

Month	Development			
	Motor	Social	Hearing and speech	Eye and hand
1	Holds head up for a few seconds.	Stops crying when picked up.	Startled by sudden sounds.	Notices bright objects when nearby.
2	Can hold head up when lying down on front.	Smiles.	Listens to rattle being shaken.	Can follow object up, down, and sideways.
3	Kicks well.	Interested and follows person with eyes.	Looks for source of sound with eyes.	Looks from one object to another.
4	Lifts head and chest when lying down on front	Smiles back at a smiling person.	Laughs.	Grasps and holds a small cube.
5	Holds head up without nodding.	Plays happily with people.	Turns head towards sound.	Pulls paper away from face.
6	Pushes up on to wrists when lying on front.	Turns head towards person talking.	Babbles or coos to voice or music.	Takes a cube from a surface.

By six weeks, the baby can control her head, unless sudden movements are made. By 12 weeks, she has complete head control, and will go on to discover other positions. She will kick if left lying on her back, and lift up head and shoulders if left lying on her front.

At about this time, a baby will learn to roll over. This can be dangerous if the baby is left alone – on a changing mat, for instance.

All children learn through play, and toy manufacturers cash in on this by producing an extensive range of toys that often overwhelms the customer into making unwise choices. With babies it is important to think of safety – there are unpleasant accounts of the results of giving babies unsuitable playthings.

Emotional needs

Parents are usually well aware of how to meet their child's physical needs – food, warmth, clothing, playthings, etc. – as there is a lot of information available about these things. New parents are offered practical guidance from the midwife, health visitor, and doctor about caring for the baby, and advertisers keep them well informed about baby foods, toys, and clothes. Amidst all this, it is very easy to forget about the baby's emotional needs. As these needs cannot be defined in the same way as physical needs can, they are not very well catalogued. The result of lack of food can be seen more easily than the result of lack of love.

The average baby can recognise her mother by three months, and by nine months will be wary of strangers. She associates satisfaction and security with the presence of her parents, which is the beginning of trust. A baby brought up in a happy and secure environment will probably grow up to be a stable and secure adult. Sadly, not all parents are able to offer this type of unselfish love.

Some parents are unable to cope with the demands of a baby. They may leave the baby to cry for feeds, and not keep her clean and comfortable. Feeds may be rushed, with no time spent in talking and cuddling.

The baby will still become bonded to her parents – after all, she knows no different – but the relationship lacks security, and the baby may grow into an unhappy and insecure child.

It is important to remember that a baby cannot be spoiled by love. In the long run, the more time and love a parent gives a child, the more pleasure the parent and child will gain. A baby who is given only limited love and care will become more demanding in the search for attention.

The baby at six to twelve months

At this stage, the baby's weight gain has slowed down, and bottle-fed babies can be put on to cows' milk, although the bottles must still be sterilised, and it is safer to boil the milk first. Cows' milk is not the complete nourishment that breast-milk is, so the baby's solid food should offer the additional nutrients.

At this stage, most babies will have dropped the early-morning or bed-time feed as they gradually become weaned. Weaning should be done carefully in order to avoid meal-times becoming a battle-ground. Sucking is satisfying and secure for a baby, so the bottle or breast should not be taken away until the baby is ready. A compromise needs to be reached as, if the mother allows the baby to wean herself, the baby may still be having a bottle by the age of four. Many babies will return to the comfort of sucking when they are ill or upset.

Breast-fed babies can be weaned straight on to a cup without using a bottle. In the same way that sucking stimulated the milk supply, the supply will diminish when the baby takes less milk.

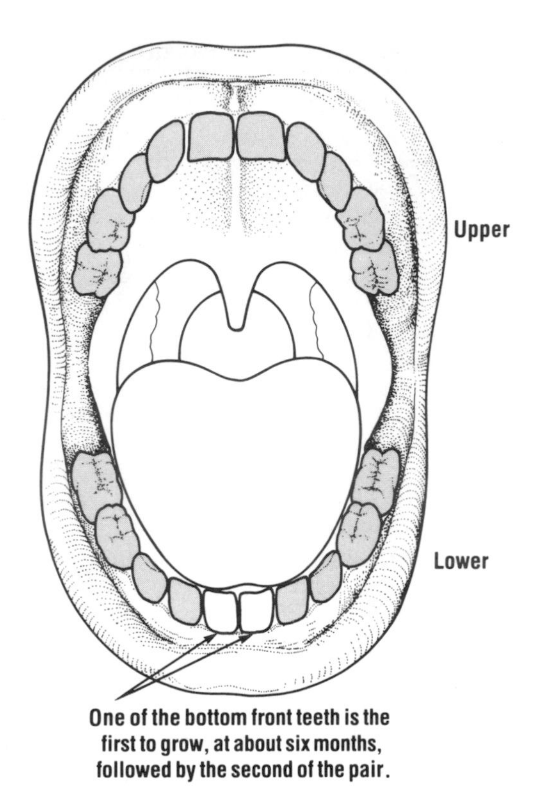

Fig. 6.2 Order of first teeth

Teething

At about six months, or even earlier, the baby will cut her first tooth. Teething tends to be blamed for many of the baby's minor ailments, and the danger here is that the symptoms of a real illness may be ignored on the grounds that the baby is teething. Teething is blamed for colds, coughs, bronchitis, and even sickness, diarrhoea, and convulsions. Although the actual teething cannot cause these illnesses, perhaps the baby is more open to infection if the discomfort makes her run-down. Some doctors do not believe that teething causes discomfort, and ask how many six- and seven-year-olds have teething problems. Teething is a controversial issue.

However, the order in which the teeth come through (fig. 6.2) is always the same, regardless of how early the baby starts teething. The first teeth cut are the two front bottom teeth. The chewing teeth do not come for some time, and the baby will quite happily chew on her gums. The baby should be given plenty of chances to chew harder foods, such as apples, carrots, rusks, etc.

To some extent, healthy teeth are inherited from the parents, but there is a lot parents can do to help keep their baby's teeth as healthy as possible. Teeth can be protected if the baby has more savoury and less sweet food. A baby can be encouraged to use a toothbrush with a little toothpaste on it in order to familiarise her with the teeth-cleaning routine. Many areas have fluoride in the water supply, and this helps prevent tooth decay.

Mobility

In the second half of the first year, babies become very mobile, so parents need to be careful about safety in the home in order to avoid accidents. Safety hazards should be considered in the house and garden, as well as when travelling or visiting. The parents should try to say 'no' only when they mean it, as the child who hears 'no' all the time may ignore it when it is important. (See chapter 22, 'Safety in and around the home'.)

Potty-training

Some parents decide to start potty-training towards the end of the first year. The baby can sit unaided and opens her bowels at a regular time each day, so it seems convenient to use the pot rather than a nappy. However, at this age the baby is not really being 'trained' as such, as she cannot recognise when she needs to use the pot. The danger of early potty-training is that it becomes an ordeal for the baby, and an eventual battleground. Research has shown that, regardless of when potty-training starts, children become 'clean' at about the same age – two and a half years.

Physical development

The baby's physical development at this age is quite amazing. Many babies go from sitting to walking in these six months, although some babies do not walk until their second year.

Babies develop their physical skills in roughly the same order and, if left to their own devices, all babies would eventually learn to sit, crawl, and walk (fig. 6.3). However, parents can do a lot to help, as babies love learning to do new things.

The baby develops muscular control from the head down. By six months, the average baby can sit supported on the floor but is not yet able to get into a sitting position unaided. The baby should be surrounded by cushions in case of a fall. It is at this stage that parents need to be particularly aware of safety; the baby should be harnessed in a pram, push-chair, or high-chair, and should never be left alone.

The next stage is crawling, which usually happens at the same time as sitting unaided. Some babies do not crawl until after their first birthday, but this delay is usually nothing to worry about. Many babies do not bother to crawl, but shuffle along on their bottoms instead. Other babies 'crawl' on hands and feet, missing out crawling on hands and knees altogether.

By about nine months, most babies are beginning to move around on the floor. Babies may get angry when they find they are moving backwards instead of forwards as they crawl, but this frustrating stage soon passes. Parents can encourage their babies during the crawling stage by dressing them in comfortable clothes that protect the knees. Although many parents find play-pens useful, they should not be overused as they restrict the baby's movements.

At about nine to ten months, the average baby will want to use her legs and will pull herself up on the cot bars or furniture. At this stage, many babies find they cannot reverse the process and sit down again. When the baby can hold this standing position and feel safe, she will move around the furniture using a sideways step. Parents should check that everything in the room is stable and can take the baby's weight. Many parents think that this is

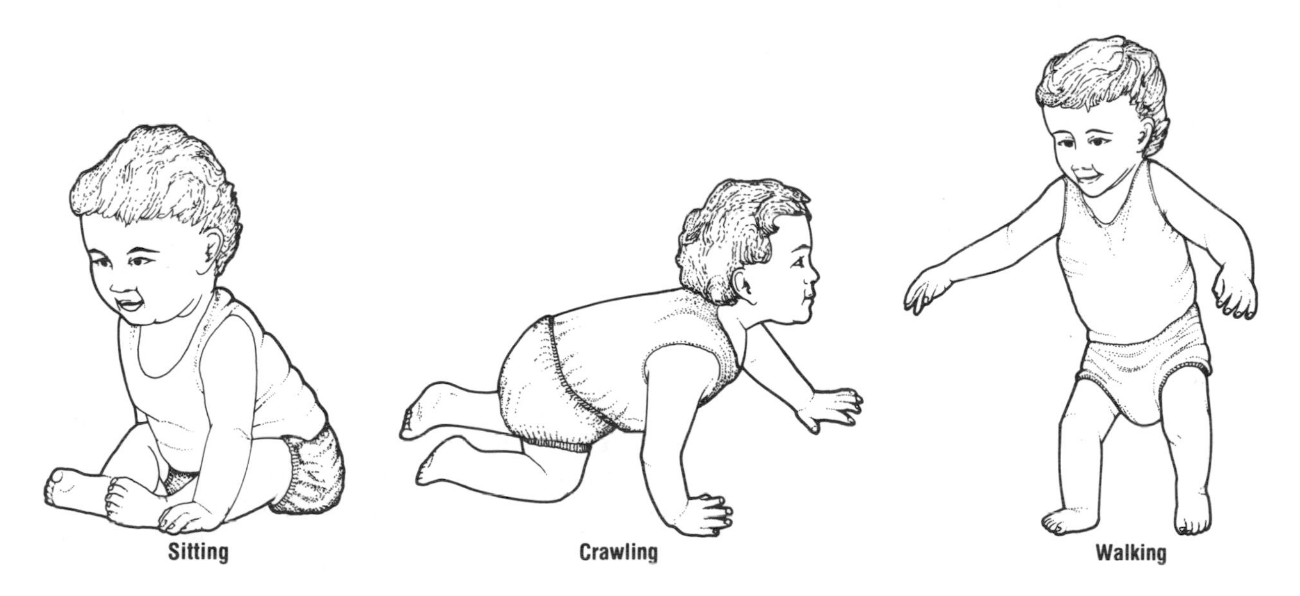

Fig. 6.3 Sitting, crawling, walking

the time to buy their baby some shoes, but ideally the baby should go barefoot for as long as possible.

Hand–eye co-ordination is a very important part of a baby's development (fig. 6.4). At about six months, the baby will reach out and grasp a toy with her whole hand and explore it by putting it in her mouth. As soon as another toy is offered, the baby will drop the first one.

At around the age of eight months, the baby can hold two things at once, and uses fingers and thumbs instead of the whole hand.

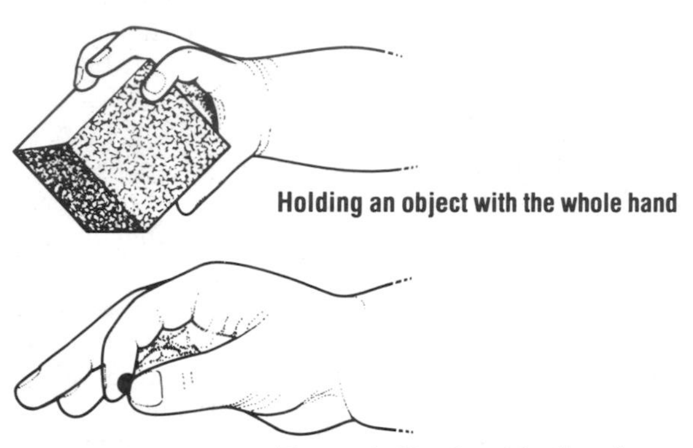

Fig. 6.4 Grips

By the age of nine months, the baby has progressed to the *pincer grip*, where thumb and forefinger are used for a delicate grip.

At around 11 months, the baby finds out about dropping objects. This is the stage where toys are dropped out of the pram or push-chair, and plates and food are dropped from the high-chair. The baby enjoys watching the object drop and hearing it fall, and later realises that people will stop to pick it up for her.

Language development

Language development is a very complex subject. A great deal has been written and there has been much discussion about how and why babies learn to talk. It is thought that babies are born with a drive to learn to speak, but, before they can do so, they need to learn to listen. Because of this, deafness delays speech development, and it is important that it should be diagnosed and treated as early as possible.

It is important that parents should talk to their baby and pick out the sounds she makes that have meaning by responding to them. The baby learns to say 'mama', 'dada', and so on, because these sounds are heard and encouraged. The reaction pleases the baby, who then tries to repeat them.

Some parents find it difficult to talk to their baby, and think it a waste of time as the baby doesn't understand. Perhaps if the parents realised the disadvantage to the child, they might be able to overcome their inhibitions.

At about eight months, the baby realises that a shout will soon bring attention, and she may also try to sing – often along with music on the radio or television. Babies enjoy the rhythmic repetition of nursery rhymes at this stage.

Towards nine to ten months, the baby's babbling begins to sound like speech, as it has the same intonations, with sounds strung together. Within the next month, it is possible to recognise certain sounds as relating to a particular object, although the sounds are not 'words' as such.

Parents notice that a baby will understand many words long before she can say them. The parents can help their baby's language development by repeating regularly used words such as 'cot', 'dinner', 'drink', 'teddy', and so on, as these words label the object. In time, the baby will learn to associate the name with the object.

Living with a baby

The baby is part of the family, and it is better for both the parents and the baby if she is included in as much of the family's everyday life as possible. Many parents find the age when the baby becomes mobile very tiresome. When the baby accidently breaks something or makes a mess, they see it as naughtiness. However, the baby cannot really be naughty. She cannot see the results of her actions and will not respond to being punished by behaving better next time.

Parents who do find this stage difficult should think of ways of reducing the problem. Tempting ornaments can be moved out of the way, the stairs can be cut off by stair-gates so that the baby will not be able to reach them, and tidying-up could be left until the baby is in bed (which would be less exhausting for the mother than running around all day trying to clear up messes that will be made again). If a more relaxed approach to child-care can be introduced, this will help the parents to cope, as they will be under less stress – and this can only benefit the baby.

Assignments

1. Introducing solid foods should be a pleasant experience for the baby. There is a wide variety of commercially produced baby foods on the market. Make a comparison between ready-made baby foods (tins and jars) and dehydrated baby foods. Look at variety, price, and quantity. How do these foods compare with home cooking?
2. What are the dangers of salt and sugar for babies? Why should they never be added to the baby's food?
3. Feeding-times need organising as the baby grows older.

   a) Arrange to visit a baby shop, or obtain a brochure, and find out what feeding equipment is on the market. Look at the choice of high-chairs; bibs; plates, bowls, and spoons; feeder beakers. When you have looked around, decide which models are best. Are they always the most expensive? Make constructive criticisms if necessary.
   b) Speak to mothers of young babies and ask them for their opinions on feeding equipment.

4. Hygiene is still important when a baby starts mixed feeding. Write a set of guidelines to be followed to ensure that the equipment and food is as hygienic as possible. Cover these points, and add any more you think are important:

   - personal hygiene
   - equipment used
   - preparation of food
   - storage of food
   - heating up food

5. 'Hyperactivity' has been discussed a great deal recently. Some people believe it is caused by food additives but, whatever the cause, the parents need help and support. Find out the symptoms and possible causes of hyperactivity, and think about the effect it would have on the parents.

   Information is available from books, and from the Hyperactive Children Support Group, 59 Meadowside, Angmering, West Sussex BN16 4BW, on receipt of a stamped addressed envelope.
6. Ask the mothers you know what they think the symptoms of teething are. Collect the information together on separate cards, and discuss each of them. Put each on a 'true' or 'false' pile. What percentage were true?
7. In pairs, find out about the development of a baby in the first year. Relate your findings to a real baby.

   a) Observe a baby of this age for 15 minutes and note what he or she does. Fill in a chart with: name, age (months and weeks), observations, comments.
   b) Find out when the average baby can: control its head, follow an object with its eyes, babble, reach out for an object, roll over, sit up (supported and unsupported), crawl, walk around the furniture, walk unaided. Is there a particular order in which these skills are acquired? Is it possible to miss out any of these stages? What should parents do if their child does not develop as expected?

8. Throughout childhood, children learn through play. Some toys are better than others at developing children's skills, and some are downright dangerous. By visiting shops and looking at catalogues and brochures, decide what toys to buy for babies of these ages: 0–3 months, 3–6 months, 6–9 months, 9–12 months.

   Choose about four toys for each age group, and say which skill you think the toy will help to develop.

   When you have made your choice, list some toys you think are bad buys, and say why.

   Finally, for each age group, think of a plaything you can make from items found around the house. The best toys do not have to be the most expensive.
9. Love and security are very important for a child. What effects do you think a lack of love and security would have on a child's future development?
10. Work out a simple weaning plan for a breast-fed or bottle-fed baby. Include these details and any others you think are necessary:

    - age of the baby at the different stages of weaning

- introducing the cup, and the best type to use
- which bottle- or breast-feeds to give up
- how to recognise when the baby is ready to give up the late-night feed
- which feeds to leave the longest
- how to avoid the baby's becoming too attached to the bottle for comfort

11. The average six month-old baby needs 800 calories a day. Work out a day's feeding programme, including milk and solid food. Milk is 360 calories per pint (634 calories per litre) and the baby has four 225 ml feeds.
12. Find out the approximate ages at which these teeth appear: lower central molars, upper incisors, lower lateral incisors, first molars, canines, second molars.
13. Find out whether your local water is treated with fluoride. If so, for how long has it been treated? If not, are there any plans to add fluoride in the future?
14. From information available from your local clinic, or from leaflets from the Health Education Council, find out about the immunisation programme by completing this chart:

Age	Vaccine	Possible side-effects and comments

15. As a group, discuss the dilemma parents have to face when deciding whether their baby should be vaccinated against whooping cough.
16. Think of ways in which parents can encourage their baby to sit unaided, crawl, move around the furniture, and stand up.
17. Babies need plenty of opportunities to learn to use their hands. List some of the ways in which parents can help their baby develop these skills.
18. Language development is a complex process. List ways in which parents can help their baby to develop listening and speaking skills.
19. Try and find an opportunity to spend some time with a baby aged between six and 12 months. Note and comment on the following, adding more if you wish:

- ability to sit, crawl, walk around the furniture, walk
- reaction when handed a toy
- reaction when handed a second toy
- how the baby grasps the toy
- level of speech development
- reaction of the baby when talked to

7 Child development from one to two years

By the age of one year, most children are nearly toddlers. They are no longer babies and are beginning to assert themselves, but they are not yet independent and they still need their parents a great deal.

Many toddlers tend to say 'no' to everything, even when they mean 'yes', and this can lead to conflict with some parents. Parents need to remember that this is not naughtiness but part of the toddler's learning to become a person in her own right, with a separate identity from the parents. Parents have to adapt to this new stage in their child's development. They have to make a balance between treating the child like a baby, which would slow down her development, and treating the child as if she were older and creating a situation the child can't cope with.

The toddler has not lived long enough to build up her experiences of life; she cannot remember from day to day whether something is dangerous or not, neither can she feel caution in a dangerous situation.

Emotions

Toddlers are not yet able to control their emotions and they tend to change moods quickly. One minute they are having a cuddle and the next a temper tantrum. Because they cannot understand concepts, it is a waste of time trying to use discipline – they are too young to be 'taught a lesson'. With some forward planning aimed at avoiding conflict situations, life with a toddler can be fun. The parents will be under less stress to prove they are in charge, and the toddler will not need to battle so much to get her own way.

Much of the toddler's so-called bad behaviour is attention seeking, which is a natural stage of development, but in a few cases it can be a sign that the child's needs are not being met. The child with this problem feels that attention given because she has been naughty is better than no attention at all.

Aggressive behaviour is another part of becoming an individual, and the child may sometimes kick, hit, or bite another child. It is rare that the toddler is doing this deliberately to hurt, as she is not aware of the results of her actions. The child who becomes very aggressive and hurts other children on a regular basis may need help to find out the cause of the behaviour.

Growth

Growth slows down in the second year, yet the body proportions are still like a baby's, with a large head and small body (fig. 7.1). As the second year passes, the child's body will change to adapt to walking on two legs rather than crawling on all fours.

Feeding

By the age of one, the child is usually eating the same food as the rest of the family, and at the same times. The foods that are good for her growth and development may seem boring and unappetising to the child, and parents should find different ways of presenting the food so that the child is still getting all the nourishment.

Some mothers become almost obsessive about the need for the child to eat everything up, and this makes meal-times a strain. The child soon senses the mother's feelings and uses meal-times as a tool against her.

Babies should not be forced to eat, and there is rarely any danger that a child will starve to death. Babies, like adults, have appetites that vary. We do not always eat the same quantities, so why should we force our babies to? Parents who try to force feeding patterns on their child may well end up with feeding problems.

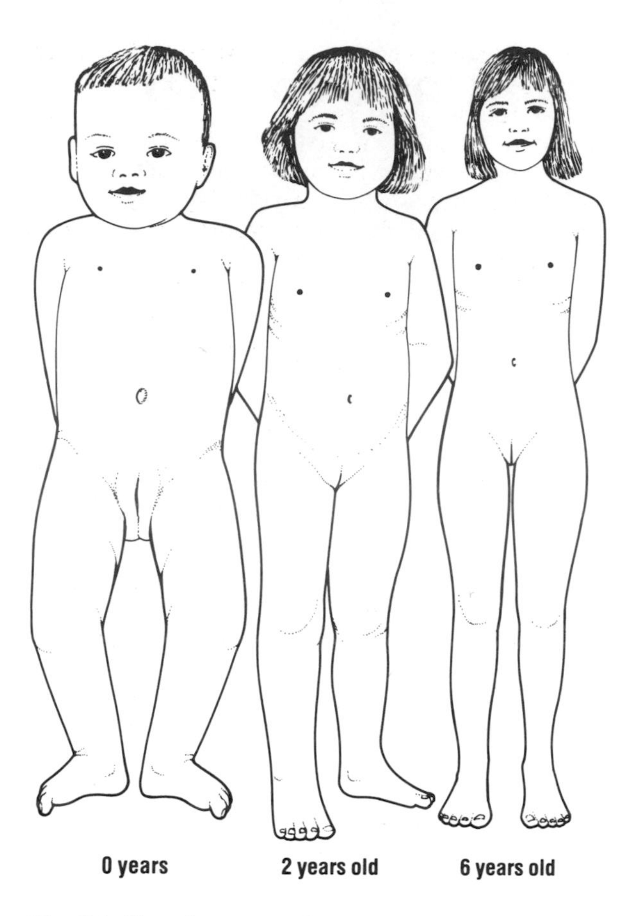

Fig. 7.1 Changing proportions

Potty-training

Another problem area for parents is toilet-training. Parents – particularly mothers – worry about when to start potty-training. It is pointless to start before the child is 18 months old as most children have no bowel or bladder control before then. At around 18 months or later, the child begins to make the association between what is produced (urine and faeces) and the sensation of needing to pass them, and it is at this stage that the potty can be introduced.

There should be no pressure attached to potty-training, as the parents' anxiety may be picked up by the child and lead to problems. Parents should praise successes and ignore mistakes, as this has been found to be the quickest method of potty-training.

Sometimes a child who has been clean and dry will revert to wetting and soiling when she is upset or ill. It is better to find the cause of the upset rather than getting angry. Often it may be the birth of a new baby in the family, for instance, or starting school. With understanding, the child should soon become clean again.

Teeth

Teething, which started at around four to six months, will continue throughout the second year. Some of the teeth may cause discomfort, but they will not cause illness. Any symptoms that develop during teething should be referred to the doctor and not put down to teething. The same good habits should continue throughout the second year to keep teeth healthy – a good diet and regular cleaning. Parents should clean the toddler's teeth as well, to make sure it is done properly.

Bones

Vitamin D and calcium, necessary for developing healthy teeth, are also needed to form healthy bones. Children's bones are fairly soft and pliable, so it is vital that they are cared for properly.

Putting a child in ill-fitting shoes will lead to damaged, and sometimes deformed, feet. A toddler should be left barefoot as often as possible.

Sleep

By the age of one, most babies have found a sleeping pattern and will sleep for 10 to 12 hours a night (and often have a day-time nap as well).

Many toddlers are unwilling to go to bed, and cry when the parent leaves the room. The toddler needs to be reassured that her parents are still there, but that it is time to go to sleep. Parents who let their child get up again find that she will expect to come downstairs again the next night and will cry until allowed to do so.

At the other extreme is the parent who leaves the child to cry, hoping to teach her to go to sleep at bed-time. This will make the child cry more, as she feels deserted, and will not help her to go to sleep.

A good bed-time routine will allow the child time to unwind after the excitement of the day, by looking at picture books, listening to lullabies, and so on. The child will feel secure in this routine and will not be so upset when the parent leaves the room. A good bed-time routine is also an advantage as it can by followed when someone else has to put the child to bed.

Caring for the toddler

It is easier to dress the one-year old as she will help by pushing her arms into sleeves and legs into trousers and socks. Dressing is a good chance for learning: parents can discuss colours, names of articles of clothing, and so on. The toddler will not be upset at being dressed and undressed any more, and may see it as a game. At other times she may be angry and frustrated at having to stay still to be dressed.

Changing nappies is very difficult, as the older baby is unwilling to lie still on her back while having a nappy changed. The parents soon learn ways of keeping their toddler occupied while quickly putting on a clean nappy, such as singing nursery rhymes, or playing peep-bo.

Most toddlers enjoy a bath, but most hate having a hair wash. Again it is a case of dealing with the problem gently and with understanding rather than forcing the child to go through an ordeal.

Physical development

The average child will learn to walk in the second year. Although no child should be forced into walking, parents can make plenty of opportunities available for the child to practise walking skills. Baby-walkers are particularly useful as they give the child some support and are designed not to tip over. In the early days of walking, the toddler will be unsteady on her feet. Many parents find reins useful at this stage, as they can be used to control direction and prevent falls. After a few months of walking, the toddler will be fairly steady on her feet and be able to walk for some distance.

Learning and play

Discovery play

Manual dexterity develops quite quickly in the second year, and by the age of 18 months, the average toddler can build a high tower of bricks. At this age, the toddler realises that certain actions bring certain results and she is able to predict these results – which is the basis of learning by experience. Through experimenting, she explores the world around her and learns about it. However, it is not until the age of two that the toddler can form concepts about her surroundings, so learning and play are still at a very simple level.

Because the toddler learns through her senses, it is essential that regular sight, hearing, and development tests are carried out to ensure that the senses are functioning properly (see chapter 11).

Learning and play are an inseparable part of the toddler's mental development. It is only later on at school that these two become separate. Meanwhile, the toys the toddler plays with can help promote learning. Toys do not need to be the expensive 'educational' type – there are plenty of cheaper alternatives in the home that are just as good – saucepans are often more fun than any toy, for example.

In this second year, the toddler begins to realise that her actions produce results: balls roll, brick towers fall over when pushed, and so on. She will also begin to differentiate between various things. For example, food is not like a toy; the cat is not a teddy; etc. This is the very beginning of forming the concepts that will be developed in the third year.

The toddler can be encouraged by being talked to and questioned about everyday life and by playing with other children, all of which broaden her awareness of the world around her.

Physical play

The type of 'discovery' learning play we have discussed develops the child's mental abilities, but children also need physical play to help them learn about their bodies. Physical play helps develop co-ordination, as well as using up excess energy when the child 'lets off steam'. Physical play can take the form of climbing, running, hopping, turning head-over-heels, and balancing. Toys such as climbing frames are available, but these are very expensive. There is usually a good alternative somewhere in the home.

Parental involvement

The toddler needs to rely on her parents to make opportunities for play by providing the playthings and the chance to play with them. This is not suggesting that the child should be allowed to play with what she likes when she likes. Some activities, such as using play-dough or painting, have to be limited to a certain time or place because of the preparation and mess involved. However, parents should put aside certain times to play with their child, as some games are impossible to play alone. Hide and seek needs at least two people, but is fun and teaches the child a great deal. There are some playthings that need to be explained or demonstrated to the child, or new ideas to be pointed out. Children learn a lot from imitation, and parents have a lot to offer their child.

Social play

Children need to be able to play with each other, and to learn how to share and how to play together. This is called 'social play'. Many toddlers go to a parent-and-toddler group with their mothers or fathers where they meet other young children and play with the toys provided. As well as giving the toddler a chance to meet other children, it prepares her for playgroup, and it gives the parents a change of scene.

Assignments on development from one to two

1. Toddlers like adult attention and will try many ways of getting themselves noticed. Think of ways in which a toddler might attract attention.
2. When a toddler feels angry and frustrated, she may have a temper tantrum. Describe a temper tantrum and think of some possible causes. How best could a parent handle a temper tantrum?
3. Many babies develop 'comfort habits' which are carried on into childhood. These habits give the child security, especially if she is unwell or miserable.
   - List as many comfort habits as you can.
   - Ask any mothers you know about their children's comfort habits, and bring your answers to the group to discuss.
4. What would be an appetising yet well-balanced day's food for the average one-year-old? Plan breakfast, lunch, and tea, and include any drinks. Give reasons for your choice.
5. Write some guidelines to avoid feeding problems. You may include any of the following, and add anything else you think is useful:
   - feeding herself
   - should meal-times be strict (manners, etc)?
   - the mess
   - meal-times as a pleasure, or something that has to be done
   - sweets in between meals, and when used as bribes
   - should the child be made to clear the plate?
6. Potty-training can be a source of anxiety for parents and children. In pairs, find out the following information:
   - What type of potty is best for the child (boy or girl)?
   - When is the best time to start potty-training?
   - Why is bowel-training easier than bladder-training?
   - Why is it wrong to force a child on to the potty?

   Write a set of guidelines for bladder-training. It may help to cover these areas: recognising the need to go, holding the urine, trainer-pants.
7. Tooth decay can be painful and result in lost teeth. Although good or bad teeth are partly inherited, there is a lot we can do to help prevent tooth decay. Find out what can be done, using these headings: diet, fluoride, visiting the dentist, sugar, brushing and dental hygiene.
8. Because children's bones are soft and pliable, it is important that their feet are cared for to prevent them from becoming damaged. Answer these questions on foot care:
   - Why is it better for a toddler to be barefoot as often as possible?
   - Why are cotton socks better than nylon?
   - Why are socks that are too small as dangerous as small shoes?
   - When it is time to buy the first pair of shoes, what advice would you give the parents?
9. Parents with wakeful toddlers are often at their wits' end over how to cope. Read the problems below and suggest some helpful ideas, pointing out any mistakes the parents may have made:
   - 'My baby throws off his bedclothes in his sleep. When I cover him up again he wakes up, and that's it.'
   - 'My baby always wakes up when I go to bed. I wondered if it was the wet nappy that woke her, so I stopped her having anything to drink after tea-time, and now the problem is worse. She sleeps in our room, so I'm getting really tired.'

- 'It's amazing, but my baby must have a sixth sense. When my husband and I go to bed and turn the lights off, she always wakes up and screams until we put them on again. What can I do – I don't want to leave all the lights on?'
- 'My toddler wakes up so early in the morning I'm getting really tired. Just one extra hour in bed would make all the difference.'

10. Finger-nails and toe-nails should be kept short, to prevent the toddler from scratching herself or other people. How should finger-nails and toe-nails be cut?
11. The toddler should be dressed in practical and comfortable clothes. In pairs, decide what clothes would be best in the following situations, for a boy and a girl:

    - everyday underwear
    - rainy weather
    - hot sunny weather
    - bed-time (summer and winter)
    - day-time (summer and winter)
    - outdoor winter wear

    If it is not possible to arrange to go and look round a baby-wear shop, then get hold of one of the many catalogues or brochures available.
12. Discuss ways that parents can encourage their child to walk.
13. By the age of two, most children have formed concepts about the world around them because of the discovery play they enjoyed as babies and toddlers. Think of some playthings – not necessarily bought from the shops – that will help the child make discoveries about the world around her. Lay your information out as a chart, e.g.:

Plaything	Discovery
Bowl of water	Makes you wet, splashes, makes a noise, leaks through fingers, can't be held, etc.

14. Physical play helps to promote the child's co-ordination and agility. Think of some playthings – again, not necessarily bought – that will help to develop these skills.

Plaything	Skill promoted
Big beach-ball	Hand–eye co-ordination needed for catching and throwing

15. Books are good playthings: they reflect everyday life and give the child an opportunity to use her imagination and listening skills. Children enjoy books long before they can read. As a group, discuss in what ways books can be valuable playthings for a toddler. Visit the children's section of your local library and make a list of five books that you feel would be good for a one- to two-year-old. What qualities did you bear in mind when making your choice?
16. An unsuitable toy is a waste of money, and could even be dangerous. Make a list of questions you should ask yourself before buying a toy.
17. Arrange a visit to a local parent-and-toddler group and look at the type of toys there. Watch the children at play and see if you can notice the different types of play – discovery, physical, and social play.

8 The pre-school years

Although the pre-school child is generally accepted to be aged between two and five, the difference between a toddler and a pre-school child cannot simply be measured in years. The difference lies in their language development and physical abilities.

In general, it should be possible to reason with pre-school children, and they should be able to communicate their feelings to other people. They are therefore more independent and less demanding of their parents' attention.

Growth

The growth rate continues to slow down, but should not stop altogether. Any child who does not grow in six months should be referred to the doctor in case there is a growth-hormone deficiency.

Feeding

By this age, there should be few feeding problems and, unless the child has a food allergy, she will be eating the same food as the rest of the family. Parents can gradually introduce table manners, so that the child knows how to eat on the more formal occasions. Parents need to remember that manners are learnt by imitation, so a child who sees a bad example will not learn how to eat nicely.

For many people, eating is closely linked with the emotions. Many overweight adults eat for comfort and to ease their feelings of unhappiness, boredom, or loneliness. These feeding associations probably started in childhood, when their mothers may have used sweets for bribery and punishment. It is quite common to hear parents saying, 'No sweets if you are naughty,' or, 'If you're good I'll buy you some sweets.' In time, the child who is treated like this will soon realise that she can use eating as a weapon against the parents. She may refuse to eat, and enjoy watching the fuss this creates.

Some mothers will insist that their children eat everything on the plate, even if it is something they genuinely dislike. This insistence does seem unfair when we, as adults, are able to exercise choice over the foods we eat. However, that is not to say that the parents should give in to the child's unfair demands about food – it is important to achieve a balance and not to create an emotional issue out of food.

It is quite common today to hear mothers express their concern about the quality of the food their children eat. They are worried about eating between meals, so-called 'junk' foods, and sweets. Again, the best approach is to strike a balance somewhere in the middle. It seems reasonable to have a mid-morning snack with a drink, and there are plenty of healthy alternatives to biscuits. Sweets need not be totally banned but can be a special treat, and the odd junk-food meal can do little harm.

Sleep

When the child is around two, most parents will be thinking about transferring her to a bed, if they have not already done so. The bed should be the child's very own special place, even if the room is shared. If the child likes her bed, there are less likely to be problems at bed-time. By this age, most children will be sleeping well, and the only disturbances to sleep will be illness and nightmares.

Children who need less sleep will wake early in the mornings. If the parents would like more sleep themselves, they should make sure their child is potted and has a drink and something to play with, and then return to their bed. If they get up when the child wakes, she will not realise that such early waking is anti-social.

Toilet-training

Toilet-training may still be carried on through the early pre-school years. Some children will be using the toilet, whereas others will still prefer the potty.

Child-sized toilet seats that fit over the normal toilet seat can help make the transition from potty to toilet easier.

Although the child is completely dry during the day at least, she will not be able to hold on for long once she feels the urge to go to the toilet. She will need to get used to using public toilets, or hedgerows if necessary. In this country, child-care is still very sexist, and very few men's facilities, either toilets or changing-rooms, cater for men with small children: it is presumed that the mother will look after them. It is common to find nappy-changing facilities in the ladies' room, and little boys with their mothers, but there are few, if any, nappy-changing rooms in the men's toilets, and they are hardly suitable for little girls who are with their fathers. Parents who would like things to change should contact their local councils.

In the past, great emphasis was placed on opening the bowels daily: someone who didn't do this was thought to be constipated and therefore unwell. Many parents' views seem to reflect this attitude, and they are unduly concerned about their child's bowel habits. Parents should only become concerned if the child complains of pain or discomfort. After all, a person who opens his bowels three time a day can still be constipated.

At this age, children do sometimes have 'accidents', and some are more prone to this than others. Accidents are more likely to happen if the child is engrossed in a game, or over-excited, and misses the signals from her bladder. Generally, those accidents will get fewer and fewer as the child gets older. Occasionally, there are problems with toilet-training, and any parents who are worried should ask their health visitor or doctor.

Children take longer to be dry at night because the signals from the full bladder may not wake a sleeping child. However, when the child seems happy to come out of nappies at night, it is a good idea to protect the mattress with a waterproof sheet so that any accidents don't matter.

Many parents 'lift' their child before they go to bed at night in the hope that this will prevent a wet bed. Although this may work in the short term, the long-term aim is to teach the child all-night bladder control, and lifting the child half-way through the night does not achieve this.

In some cases, a child who has been dry at night will start to wet the bed. As this is often caused by stress, the parents should find out the reason for this backward step and try not to be cross as this may make matters worse. The child who starts soiling again should be referred to the health visitor, as this is usually a sign of a deeper disturbance.

Play

The average pre-school child is well able to control her body as she has discovered her strengths and weaknesses through play. Playing will prepare her for her future life at playgroup and school, and, by the age of two, most children have learnt some concepts about everyday life through the games they play. Once the child has reached this stage she can enjoy creative play, where she can make things, using her imagination.

Creative and imaginative play

Creative play is often messy, and some parents may try and dissuade their child from creative play because of the extra work involved for them. This attitude will put their child at a disadvantage later on in playgroup and school, where creative play forms the basis of much learning. An example of this is the child who learns her colours and can hold a pencil before going to school: this child is going to be further ahead than the child who has not mastered these skills.

Imaginative play, or 'let's pretend', usually develops at around the same age as creative skills appear. The child is able to use her concepts of everyday life to stimulate the imagination and make up games. She might play hospitals, or mummies and daddies. The games may not seem realistic to the outsider, but imagination makes them real to her.

Nowadays, there are so many toys around that there is a danger that nothing will be left to the imagination; but in the past the same was said of television. It seems more likely that the child's imaginative abilities will not be swamped, regardless of anything produced by adults for children: children do not need elaborate 'props' to make their games realistic.

Pre-school groups

The pre-school child is gradually learning independence: she is fairly able to look after herself, and to ask for help if necessary. She is forever asking questions about the world, and is probably ready to mix with people from outside the home more, either at playgroup or nursery school. It is no criticism of the home to recommend that most children would benefit from going to one of these – they offer better opportunities for mixing with other children of the same age, as well as offering a wide variety of play experiences. Pre-school playgroups or nurseries are valuable to all children, but especially to children who are disadvantaged in some way.

Most of these groups will take children at two and a half years old if they are ready to leave the mother, and by the age of three most children will be ready to benefit from a little independence. The groups do insist that the child is able to use the toilet alone, as they do not have the staff to cope with nappy-changing.

The pre-school child may attend one of three pre-school groups:

1. *Day nurseries* These aim to care for children who may need more care than they receive at home, or whose parents go to work. As their main role is the physical day-to-day care of the child, less time may be spent on the child's educational development.
2. *Nursery schools* Nursery schools like to see themselves as playing an educational role in preparing the child for school. Some of these schools are privately run and some are funded by the local authority. Most of the staff are fully-trained nursery teachers who know how to bring the best out of each child.
3. *Playgroups* Playgroups try to give the child plenty of enjoyable play and learning experiences before she starts school. They are usually run by local mothers on a committee basis, and supervisors are employed for a small fee. A small charge is made for each child to cover costs, but most of the money comes from fund-raising. Playgroups encourage parental involvement, as they see their role as extending what the child learns at home, rather than taking over from the parents.

The first few sessions at any of these pre-school groups are important. If the parents take time to prepare the child for the new experience, she is more likely to enjoy the time spent there and benefit from what is on offer. The child who is not prepared, and feels abandoned by her mother, will not settle down for some time.

Discipline

Many parents are concerned about discipline, and are frightened of 'spoiling' their child by giving her her own way. We need to ask ourselves what we mean by discipline – do we mean 'blind obedience' or do we mean the type of discipline which prompts the child to do what she knows to be right without having to be told what to do?

It is to be hoped that most parents would prefer the latter, where the child learns to govern her behaviour in accordance with guidance given by the parents in the early years.

Discipline is really a case of give and take: parents cannot expect respect from a child who is never shown respect by them. Adults need to know why they have to do certain things, and so do children. They are more likely to think next time if they are told why they should not leave the gate open or pull the cat's tail, as they will realise the results of their actions.

A child who is over-disciplined is less likely to respond to discipline, because frowning, shouting, and punishment are likely to be a large part of her everyday life. The child who is used to having friendly and sympathetic parents is more likely to respond to a cross face and therefore less likely to need a smack.

A pre-school child is often naughty in order to test her parents and see how far she can go. This can be quite a battle at times, and the parents need to approach the situation carefully, perhaps by following these guidelines:

- They should try to be consistently fair when teaching the child how to behave. The child needs to know that 'no' means 'no', and no amount of arguing and whining will change the parents' minds. If parents who say 'no' finally give in and say 'yes', then the child will remember this and repeat the excercise again until she gets her own way again.
- The child needs to know where she stands: behaviour that is accepted one day should not be punished the next. If the child gets this sort of poor discipline at home, she will not know how to behave and will suffer later at school, or when playing with friends.
- Parents who always use smacks and threats to discipline their child may have a short-term answer, but in the long run the child will not learn how to behave. Smacking is often the

easy way out, used to relieve the parents' own anger and frustration.

The child who is disciplined by firm, kind, and consistent parents is more likely to feel happy, safe, and secure. She will relate to other people better, and develop her own self-discipline. The undisciplined child may be spoilt and unpopular because of her selfishness and rudeness. The child who is over-disciplined will be unhappy and miserable, and may well bear a grudge against her parents in later life. Worse than this, she may repeat the same pattern of over-discipline with her own children.

Language development

As language development progresses throughout this pre-school stage, the child uses language to communicate ideas as well as her immediate physical needs.

Parents can help their child's language development enormously by holding frequent conversations with her. They should take any opportunity to increase the child's vocabulary by comparing sizes, heights, colours, numbers, and so on, to show the relationships between words.

Many parents remember this age as being full of 'why?' questions. Each time, the parents' answer is followed by yet another 'why?'. The child is not doing this to annoy them, but to find out about things as her curiosity increases.

Slow speech development may cause concern at this stage, so any child who seems slow in learning to speak should be referred to the doctor or health visitor. The problem will probably be picked up at the baby clinic during one of the assessment checks.

There are many reasons for slow speech development:

- Sometimes it can be inherited from one of the parents. It is a good idea for worried parents to ask their parents when they learned to talk.
- Children need language to imitate before they can learn to talk. The child who does not hear enough language from her parents is likely to be late in talking. Pre-school groups can be valuable in this situation by helping to stimulate the child's speech, and that is why children whose speech is slow in developing are given priority.
- Deafness is one of the more serious causes of slow speech development, because the child who cannot hear will not be able to decipher what other people are saying, and will not even hear her own babbling. Without hearing, the child will not learn to speak without special help. (Deafness is covered later, in chapter 17.)

Assignments on the pre-school years

1. A growth-hormone deficiency may prevent a child from growing. Nowadays, this condition can be treated if diagnosed early enough. However, out of ten children referred to a specialist, only one actually has an abnormality. In pairs, find out this information on growth:
   - What is 'normal' height growth, and how is this worked out?
   - What hormone controls growth?
   - Why does the specialist ask about the relationship between the parents and the child when considering growth problems?
2. Food allergies are becoming more common, but researchers are not sure whether this is because of better diagnosing or because the food we eat nowadays is more processed.
   - What types of food are likely to cause allergic reactions?
   - What are some of the reactions shown in food allergies?
   - Are any members of your group allergic to certain foods?
   - If a mother thinks her child is allergic to a certain food, what should she do?
3. Spend a few minutes thinking about your eating patterns and ask yourself these questions:
   - Do I eat more when I am bored/unhappy/lonely, or do I eat less?
   - Do I feel overweight?
   - Do other people see me as overweight?
   - Would I like to be slimmer?

   Do you think your attitude to food has been affected by factors in your childhood?

   As a group, discuss how parents can help foster healthy eating habits. Note down any useful comments.
4. Orange squash and biscuits are not particularly good or nourishing for the child. Think of some better mid-morning snacks.
5. Many convenience foods are a good source of nutrition, and others have very little food value

at all. Make a chart listing ten convenience foods, and comment on their suitability for a young child.

6. Some parents attempt to impose a total ban on sweets, whereas others seem to allow free access to the sweets' tin. As a group, discuss a sensible approach to eating sweets. Include some of these points if you wish:
   - the possible results of taking the two extreme attitudes mentioned above
   - how to develop a healthy attitude towards sweets
   - when to eat sweets
7. In small groups, discuss the dangers of forcing a child to eat everything on her plate.
8. Suggest a list of food alternatives that can be used when children dislike certain foods:
   - Fish and meat are the most common sources of protein, but not all children like them. What alternative sources of protein are there?
   - 'Eat up your cabbage, it's good for you.' Although cabbage is a source of fibre and vitamins, it is not a particularly good one as cooking destroys much of the food value. What other sources of fibre could the child eat?
   - Fresh fruit is a good source of vitamins. In what other ways can a child get vitamins (apart from taking vitamin pills)?
9. In extreme cases of an unbalanced diet, the child may suffer from a number of ailments. Complete this chart:

Disease	Symptoms	Caused by shortage of:

10. Plan a bedroom for a pre-school child. Remember to include bed and bedding, sleeping clothes, toys and playthings, storage for clothes and playthings and books, things for the wall, furniture.

    You could cut the things you want for the bedroom out of a catalogue, and make a collage.
11. Using mail-order catalogues, brochures, and/or visits to baby-care shops, find out what is available for toilet-training children. Some of these items will be useful, but many will be a waste of money. Make a chart of the items, together with your comments and criticisms.

Item	Price	Comment	Criticisms

12. Find out whether any of these men's facilities in your area cater for the care of young children: public lavatories, swimming-pool changing-rooms, store toilets, station toilets.

    If the answer to any of these is no, did they offer child-care facilities for women?

    Do you think there should be unisex toilet and changing facilities available as is the case on the continent?
13. Creative play, which is an essential part of a child's development, can teach:
    - colour discrimination
    - hand–eye co-ordination (by using scissors, etc.)
    - using and holding pencils, crayons, paintbrushes, etc.
    - moulding shapes
    - measuring and comparing shapes, sizes and quantities, etc.
    - rhythm

    Think of some creative-play situations that can help to develop these skills.
14. Imaginative play gives the child an opportunity to learn about other people and situations through her own fantasy world. As a group, discuss how imaginative play could help a child cope with situations in her everyday life.
15. Go to your local library and choose a book you think a pre-school child would enjoy. Read the story to a child and note his or her reactions. (If you have difficulty finding a child of this age, the local playgroup would probably be willing to help.)

    After you have read the story, ask yourself these questions:
    - Why did you choose the book?
    - Did the child like the book for the same reasons?
    - Did the child enjoy the story as you were reading it?
    - Were there enough illustrations?
    - Did the child like the illustrations?
    - Did you still feel it was a good choice after reading it?

16. List any home circumstances that suggest that a child might benefit from going to a pre-school group. These need not only be problem situations.
17. In pairs, find out about the pre-school provision in your area. Include nursery schools or classes, playgroups, day nurseries, and day centres. To find out about each, use these questions, adding more if you wish:
    - Who runs the group?
    - Are the staff trained or untrained?
    - What does it cost?
    - How are places given?
    - What are the aims and intentions of the group?
    - How often does a child attend?
    - Does it operate all year, or termly?
    - How long is each session?
18. Individually, or in small groups, try and arrange a visit to any pre-school group in your area to observe a typical session. At your visit, look for the following:
    - Is the group happy for you to visit?
    - Do the staff seem interested in the children?
    - Are the children happy?
    - Are the children kept occupied in different tasks?
    - Are the children well supervised?
    - Is there a variety of the different types of play: quiet play, creative play, discovery play, physical play?
    - Are the surroundings pleasant and bright so as to stimulate the children?
    - How large are the small groups?
    - Is there a break?
    - Are there any 'mother helpers' or parental helpers?
19. Starting a pre-school group can be a traumatic experience for a three-year-old child. As a group, list ways that a parent can make the child's first days at the group as pleasant an experience as possible.
20. Discuss what you think discipline is, and include some of these points:
    - Is discipline at home different from discipline at school, college, or work?
    - Should we encourage children to question discipline by giving them reasons for certain orders?
    - Should parents say, 'Because I said so,' if their child asks why she has to do something?
    - How were you disciplined at home?
21. It is thought that praise may be a better way of disciplining a child than punishment. However, it is sometimes necessary to punish a child as a last resort. In small groups, discuss these questions:
    - In what situations would praise be the best type of discipline?
    - Why does a child need to be old enough to understand why she is being punished?
    - What are the best ways of punishing these acts: biting another child, drawing in a book?
    - What are the dangers of punishing too long after the event – e.g. mothers who say, 'Wait till your father gets home!'?
    - Why is physical punishment not the best way of disciplining a child?
    - Why should parents not make empty threats?
    - Say which of the following should be the purpose of punishment and explain why:
        - to ridicule the child
        - to relieve the parents' anger
        - to teach the child respect
        - to show the child the results of her actions
22. The pre-school child uses language to communicate ideas and to ask questions. In pairs, find a pre-school child you can spend some time with. Talk to the child and record the conversation on tape.

    Write a report on your findings, asking yourselves these questions:
    - What is the child's age and sex?
    - Does he or she use pronouns (you, me, I, etc.)?
    - Can the child pronounce all sounds correctly? If not, which sounds does he or she find difficult?
    - Does the child stammer at all? If so, on which sounds?
    - Can the child carry on a simple conversation?
    - Is it easy to understand what he or she is saying?

9 Starting school

In the United Kingdom, it is compulsory for a child to attend full-time education by the age of five.

Starting school is a major event in a child's life, as it is probably the first time she has left home for any length of time on a regular basis. For the first time, the child will have to:

- mix with a large number of children, and no one knows whether the child will mix well and be accepted, or be bullied and unhappy. In most cases, it is the parents who worry about this: the child remains blissfully unaware of any problems at this stage.
- cope with authority and discipline, and adapt to a fairly rigid timetable. She will have to learn to obey all the different teachers as well as the headteacher, the dinner ladies, play assistants, and so on.
- become more independent, as the teacher will not have that much time to give each child, especially now that classes are getting larger in some areas.

As we have said in chapter 8, pre-school groups are a good way of preparing the child for school. The group helps her to socialise with other children and get used to being away from home. However, all the good that a pre-school group does to prepare a child for school is useless if the parents have negative attitudes towards education. Parents who do not value school will pass their negative attitudes on to their child; parents who value school will pass on more positive attitudes, and their child will be happier to start school.

Ideally, the child should be made aware of school well before it is time to start. School can be introduced through discussion, books, and television, and this will help make the transition from pre-school to school easier.

The school

Today, most reception classes are very friendly places and the children are made to feel welcome. The teachers are aware of the child's feelings when starting school, and will do what they can to help the child settle in happily.

A child will face many new situations at school. The day will probably be more structured than at home or in pre-school groups, with set times for breaks and lunch, set days for PE, and so on.

Classes in many schools are very large, so children need to compete for the teacher's attention, and some may get left out. Good teachers will try to give all the children some attention, but not all teachers have the skill to do this. The child with a secure and stable home background will be more able to cope with the challenges of school than the child who is not yet emotionally ready for school. The teacher should be responsible for noticing the child with problems.

It is quite common for children to develop the occasional problem at school. Some have Monday-morning aches and pains, or worries over a certain teacher. Others may be bullied. Parents and teachers need to be ready to spot any areas of difficulty before they develop into deeper problems.

At best, school can be an outlet for many of a child's needs – it gives the opportunity to make friends, to learn, and to become more independent. At worst, it can be the place where a child's self-confidence gets battered.

Awareness of the body

As the child grows older, she will start to ask questions about her body. The child is not being rude, so parents should try not to make the child feel guilty, and answer her questions truthfully. Although the information is best coming from the parents, there are books and pamphlets available that will help if they do find this impossible.

Play

Children at primary school still learn through play in much the same way as do pre-school children.

They can be creative in drawing, painting, music, and other arts and crafts, both at school and at home. By this age, the child knows what she intends to draw before starting the picture.

Language development

Language development continues to be helped by the child's looking at books and pictures and listening to stories. Children still need to converse with adults who can ask the right sort of questions which will stimulate their imaginations.

Feelings

Children learn how to cope with their feelings through playing with other children. They discover how it feels to be angry, jealous, or frightened, and find out how to come to terms with these feelings. They also become competitive, and try to be the best at everything they do. School encourages this competitive side of the child's nature in class, in PE, and in the playground.

Discipline

Disciplining the older child depends very much on the type of relationship she has formed with her parents in the earlier years. If there are existing discipline problems, they are unlikely to disappear but may become more apparent as the child is disciplined by people other than her parents.

Assignments on starting school

1. Try to arrange a visit to your local primary school to find out about the reception class. On your visit, ask if the school does any of the following to help the new child settle in:

   - arrange for pre-school groups to visit during the term before the child is due to start school
   - have half days at school for the first few weeks of the child's first term
   - let the mother stay for a while if necessary
   - invite the parents to look around before their child starts school

   While you are in the reception class, note the type of activities available that will encourage the children's development.
2. What manual skills would it be useful for the child to have achieved before she starts school?
3. What comments do you have to make on these mothers' attitudes to school:

   - 'There's your teacher. She'll hit you if you're naughty, so you'd better be good.'
   - 'I can't wait for you to start school – it's about time I had some time to myself.'

4. Obtain some copies of the Health Education Council leaflet 'How we grow up' (SE 2).

   - What age group do you think the leaflet is intended for?
   - What is your opinion of the leaflet?

5. Parents may worry about whether their child is 'doing well' at school. Obviously, children have different abilities, but sometimes there can be a problem that prevents a child from achieving at school. Check that you understand the symptoms of these disabilities and discuss any problems they may cause the child at school, and ways in which the child could be helped:

   - dyslexia
   - problems at home
   - minor handicaps in sight and hearing

10 Care of the disadvantaged child, fostering, and adoption

A child is socially disadvantaged when she lives in an environment which is likely to prevent her from growing and developing at a normal rate. There can be a number of reasons for a child's being socially disadvantaged, but it would be wrong to think that these factors always lead to problems: the majority of children in these situations grow up into normal adults. Some of the factors in question are:

- a working mother, or a one-parent family
- poor housing conditions
- poverty
- poor parenting – often through ignorance of child-care methods, or lack of intelligence
- a depressed mother, either through mental illness or poor home conditions

It is important for the child that any problems should be picked up as soon as possible, so that steps can be taken to remedy the situation. The people most likely to pick up these problems are health visitors, social workers, and teachers, as they are in contact with the families on a regular basis.

In the long run, perhaps improved housing conditions and better education may reduce the number of disadvantaged families. Meanwhile, while there is a problem, local authorities have to do something to help.

Foster care

Many local authorities provide day care and full-time care facilities for children whose parents are unable to look after them satisfactorily. In cases where there is a long-term problem, the social services may suggest that the child is fostered.

Foster parents will care for a child on a short- or long-term basis (long-term is more than eight weeks), and for this they receive an allowance from the local authority to cover expenses. The foster child is under the supervision of a social worker, who will check her progress and support the foster parents.

Adoption

Adoption gives a child a permanent home when her parent(s) are unable to do so. The adoptive parents have complete legal rights over the child, and can bring her up as their own.

Adopting a child is a fairly lengthy process, involving counselling between the social worker and the adoptive parents, and there are legal processes to be completed.

Assignments on the disadvantaged child, fostering, and adoption

1. In small discussion groups, take one of the following causes of social disadvantage and decide how it can create an unhappy environment for the child:
   - a working mother and/or one-parent family
   - poor housing conditions
   - poverty
   - poor parenting
2. Day and full-time care facilities for disadvantaged children are provided by the local authority. Find out about the provision of each of these in your area:
   - day nurseries
   - local-authority care
   - community homes

   Find out what each of these facilities offers the child and how it is run. Perhaps you could arrange for someone who works in one of these places to come and talk to the group about his or her work.
3. A child in difficult circumstance may be put in a foster home on a short- or long-term basis. Find out the following information:
   - How does your local social-services department recruit foster parents?
   - What qualities does it look for in foster

parents?

- What difficulties and problems may be faced by: the foster parents, the foster child?

4. Find out if there is a foster parent in your area who would be willing to come and talk to the group about the pleasures and problems of fostering. Perhaps a social worker would be willing to refer you to someone who could help.
5. Adopting a child is becoming increasingly difficult, as there are fewer children available for adoption. The process of adoption is quite lengthy and is intended to make sure that the adoption is the right step for everyone concerned. Find out about these aspects of adoption:

- What does the law say about: the adopting parents, the child, consent for adoption, the adoption process, the court order?
- An Act passed in 1976, called 'Information of origins', allowed an adopted child to find out about his or her birth. Look up the Act and find out what it says.
- Find out about adoption agencies.
- What happens after the child has been adopted? When and how do you think the new parents should tell the child about the adoption?

One-parent families

A one-parent family is one in which there is only one parent to take care of the entire family. About one in eight families today are in this situation. More often it is the mother who looks after the family alone, but there are an increasing number of fathers taking on the care of the family. One-parent families are often highlighted as being one of the causes behind socially disadvantaged children. This may be so statistically, but it would be wrong to assume that the children of all one-parent families are deprived or disadvantaged in any way. The needs of one-parent families vary depending on the reasons for there being a missing parent.

The single parent has a difficult role to play, as he or she has to be both mother and father to the child. Many feel isolated, and feel they are having to cope with their problems alone. Social workers and health visitors are available to advise families in this situation, and there are local and national agencies offering help and support.

Many local authorities will give the children of one-parent families priority in nursery schools. This allows the parent to go to work so that poverty is not added to the family's problems.

The working mother

Nowadays, it is not unusual for mothers of young families to go to work. This can be for a number of reasons: the family may need the money because of unemployment or a low wage, or because it is a one-parent family. It is also possible that the woman may choose to work rather than be at home all day, as she may have had a successful career before having her family. Whatever the reason for the mothers' working, she will need to make child-care arrangements.

In addition to this problem, she may also have to cope with a husband who does not want her to go out to work, or who is unwilling to take his share of running the home.

The working mother will also have to expect feelings of guilt now and again as she wonders whether she is doing the best thing for her family. However, this guilt can be reduced if she feels that her children are being cared for properly.

Child-care arrangements

Child-minders These are the cheapest and most common means of looking after a child. Ideally, they should be registered by the local-authority social-services department, which has a list of child-minders in the area. However, if the child is over five, the child-minder does not have to be registered, so the mother must be responsible for choosing a reliable minder.

Day nurseries These are the ideal way of caring for the child of a working mother. They are run by the social-services department, and the staff are trained in child-care. Sadly, there are too few places to cope with the demand, and it is unlikely that there will be any plans to improve this situation in the future.

Crèche facilities Some employers offer crèche facilities for their employees' children, but, again, this is a rare thing. Crèches are convenient because the mother can go and visit her child during the lunch hour, and is near at hand should there be any problems.

Nannies Better-off families can afford to pay a live-in nanny or mother's help to look after the children, and often to do the housework and cooking as well. Very few families can afford this option.

Assignments on one-parent families and working mothers

1. What reasons could there be for a family being cared for by one parent?
2. Find out about any national organisations that help single-parent families. Write to them, enclosing a stamped addressed envelope, asking for information about the services they offer.
3. Are there any local groups in your area for single-parent families? If so, what do they offer?
4. There are a number of benefits available to one-parent families. Go to your local DHSS office, or post office, and pick up the relevant leaflets.
5. A single parent on a low income may receive supplementary benefit. If the mother has a boy-friend living with her, or the father a girlfriend, the benefit may be threatened. Find out about the position of a woman who is in receipt of supplementary benefit. Discuss your findings.
6. Discuss the saying: 'A woman's place is in the home.'
7. Find out about child-minders:
   - What is a registered child-minder?
   - What standards does she have to conform to in order to get a certificate?
   - What should a parent look for in a child-minder?
   - What qualities should a good child-minder have?

   Try to arrange for a child-minder to come and talk to the group about her day. Alternatively, a visit could be arranged to see the child-minder in her home. Your social-services department could put you in touch with a good child-minder.
8. Find out where the nearest day nursery is in your area and try to arrange to visit it. If this is not possible, perhaps one of the staff would be willing to talk to the group about his or her work.

   Apart from finding out about the everyday routine, also find out:
   - the opening hours
   - the fees
   - whether the places are limited to families in need
   - whether it is privately owned, or run by the social services

Child abuse

Child abuse is often called *'non-accidental injury'* – a term used to describe an injury to a child that is not the result of an accident.

In the UK, children are considered to have certain rights, and one of these is that parents are not permitted to harm them in any way. For this reason, child abuse is not tolerated and is dealt with by the social services and the police.

Child abuse does not simply mean baby-battering: any one of these conditions comes under the heading of child abuse:

- physical injury, or baby-battering
- emotional neglect, which may lead to mal-adjustment and behavioural problems, and is often coupled with
- physical neglect, where the child does not receive proper care. She may be poorly fed, unclean, and left alone for long periods of time. These are often linked together under the term 'physical cruelty'.
- sexual abuse, where the child is sexually abused by members of the family. This area of abuse is causing increased concern, as cases are becoming more frequent.

Most parents do manage to bring their children up in a reasonable environment, but some families seem unable to cope with the demands of bringing up a child. Research has shown that the following parents are more likely to abuse their children:

- Parents who are young, with two or more young children close in age. Most at risk are children under three years old.
- Parents who were abused themselves as children will be less likely to be able to form loving relationships with their own children.
- Parents who are from social classes 4 and 5. Although this is not always the case, it is more likely to be so, because there may well be poverty, poor housing, and other problems present.

Professionals dealing with child abuse
Professional people dealing with the family need to be aware of the conditions leading to the child abuse.

Health visitors visit many families and may well be the first to recognise a family who is at risk. The parents will need support, help, and advice, and referral to a social worker will mean extra help, especially if the family has housing and financial problems.

Most local authorities will have a review committee, made up of representatives from the health services, police, social services, education and probation services, and the NSPCC, whose aim is to co-ordinate work done with the 'at-risk' families, and check any non-accidental injuries referred to it. It will also keep known and suspected cases of non-accidental injury on an 'at-risk' register, which is regularly reviewed and updated. This is used by professionals who come in contact with the families and may need the information.

People who deal with children as part of their work – such as nurses, teachers, and so on – need to be aware of the signs to look for in non-accidental injuries. These are:

- bruises, particularly in places which suggest that the child has been shaken, or hit around the face
- broken bones, and X-rays which show unreported fractures that have already healed
- failure to thrive, and severe nappy rash
- burns and scalds, particularly burns that could be caused by cigarette ends

The abused child
Once non-accidental injury is suspected, the child should be thoroughly examined. The first priority is to treat the child and put her somewhere safe to prevent further injury. This may be hospital, or a foster or residential home. If the parent refuses to allow this, then the case may be referred to a magistrate who will make a 'place of safety' order.

As soon as the child is in a place of safety, a case conference is held in order to assess the situation. The conference, which is attended by all the professionals dealing with the case, has to decide on a keyworker to co-ordinate a programme of help for the family. Although the police can attend the conference, most child-abuse co-ordinators prefer not to prosecute, but to try to deal with the causes of the problem.

The conference may decide to send the child home under the supervision of the keyworker. In more severe cases, the child is given legal protection, under either a care order or a supervision order.

Research has helped to find out some of the causes of child abuse, but there is still a great deal of work to be done, both for the families at risk and in society as a whole.

Sexual abuse
Sexual abuse of children has been causing widespread concern recently, as the number of reported cases is rising. Sexual abuse includes incest, and other forms of sexual molestation.

The abuse itself has a damaging effect on the child, but so also do the various investigations that have to be carried out in order to establish whether full intercourse has taken place or not. Many children have to give evidence in court against a parent, relative, or friend. Recent work in the USA has found alternative methods of dealing with sex abuse that aim to help bring the family together rather than force it apart by legal proceedings. Children suffering from the trauma of sexual abuse have been helped to overcome their fears by re-enacting the scenes using realistic dolls.

Recent research has shown that as many as one in ten children in the UK are sexually abused at some time. Of these, the majority found it an unpleasant experience, and one in twenty considered themselves to have been badly affected by it. In most cases the child knew the abuser, who was either a member of the family, a friend, a neighbour or a teacher.

Although most of the abusers were male, females were responsible for one in twenty cases of child sex abuse. The vast majority of the victims were girls.

Assignments on child abuse

1. Child abuse is a very emotive issue. Many people feel that the abuser should be heavily punished, while others think society is at fault because of bad housing, poverty, lack of education, and so on. As a group, discuss your views on child abuse.
2. *Case study*

   'Tracy is 18 now. When she was younger, she went to a special school. At 15, she became pregnant by Tommy. At 16, they got married and the local authority housed them in a small flat.

Rachel was born a month later, and Darren 10 months after Rachel. Darren was always a miserable baby – he cried much of the day and refused to settle at night. Tommy was unemployed, but spent most of his time out of the house. Tracy couldn't cope with Darren, especially now that she was pregnant again.

The social worker visited regularly, but Tracy couldn't communicate her problems well. The social worker suggested putting Darren on solid foods to settle him at night, and suggested that Tracy went to a mother-and-toddler club to get her out of the house more.

One evening, Rachel had been ill with a tummy upset and Darren was crying. 'I'm going out,' said Tommy. Darren continued crying until his screaming woke Rachel. Tracy snapped. She picked up the baby and shook him violently until she couldn't shake him any more. After that he seemed to sleep.

The next day, Tracy thought Darren must be catching Rachel's bug as he didn't cry or eat.

Two days later, the social worker called for her visit and noticed that the baby's head was swollen and his eyes were bloodshot. Darren was admitted to hospital suffering from a brain haemorrhage.

Darren is now mentally retarded and has little chance of developing above the mental age of three.'

After reading the case study, answer these questions:

- List all the factors that led to Tracy's battering Darren.
- What could Tommy have done to help?
- What do you think of people like Tracy and Tommy?
- Do you think their other children are at risk?

3. The National Society for the Prevention of Cruelty to Children is a voluntary organisation providing round-the-clock help for children in need. It offers its services to both parents and children. Find out about the NSPCC and any other voluntary organisations that help families at risk.
4. A case conference may decide that the child can return home under the supervision of a key-worker who will try to relieve the causes of the situation by helping the parents to form a better relationship with their child. The keyworker will also help with any social problems that the family may have.

   As a group, discuss the possible advantages and disadvantages of this approach.
5. If the case conference decides to take the case to a juvenile court, the child may be placed under a care order or a supervision order by the court. Find out about these two orders.
6. Recently, many tragic cases of child abuse have been reported in the news, and some of these cases have resulted in the child's death. Try to obtain one of these reports and analyse the factors leading up to the injuries. Working individually, study the report and consider the following factors:

   - the approach of the social services
   - the parent(s)
   - the child
   - the home environment

   Finally, as a group, discuss your findings and compile a list of factors that lead to a child being at risk.
7. In small group, discuss sexual abuse, and include the following in your discussion:

   - Do you feel it is worse when the child knows the abuser?
   - Does the age of the child affect the situation?
   - How do you think sexual abuse should be dealt with?
   - What do you think the long-term effects on the child may be?

   These addresses may be helpful, but remember to include a stamped addressed envelope for a reply:

   - NSPCC, National Headquarters, 67 Saffron Hill, London EC1N 8RS
   - British Association for the Study and Prevention of Child Abuse and Neglect, c/o John Pickett, Jacob Wright's Children's Centre, Whitworth Road, Rochdale, Lancs OL12 6EP

11 Health checks for children

From the foetal stage and on throughout childhood, health checks are vital in order to assess the child's development. In most cases, these tests confirm that the child is normal, but occasionally they point to problems that need attention.

In the early years, a child's healthy development may be measured in terms of growth. A child who does not grow at a 'normal' rate (and what is normal covers a wide range) may be showing signs of what is called *'failure to thrive'*.

The cause may lie within the child (such as health or hereditary factors which inhibit the child's growth) or outside the child (such as a poor diet or an unstable home background).

At the other extreme is the child who is overweight, or 'obese'. Again, factors within the child may be the cause (such as gland malfunctions or inherited problems), or factors outside the child (such as a poor diet or an unstable home background) can lead to overweight as well as underweight children.

Health checks in the community are carried out by what is called the *primary health-care team*, which is usually made up of doctors, health visitors, district nurses, community midwives, school nurses, and social workers. They are called a 'primary' team because they are the first contact point with patients. If the person needs treatment, he or she will then be referred on to the *secondary health-care team*, which can offer specialist help.

It is the role of the primary health-care team to look after people's health from before birth to old age. The members work as a team in order to use time and resources as efficiently as possible and to share information about clients and families.

The new-born baby

At birth, or soon after, babies undergo certain tests to check that they are physically healthy. It is essential that any health problems are detected early, so that they can be treated.

Babies are screened for various conditions that may be present at birth. These are congenital hip dislocation (covered in chapter 17) and phenylketonuria (PKU), which is covered in the assignment section. A physical examination will also show whether the baby's bone structure is normal, and that the heart and circulation are in good working order. The abdomen will be checked for any abnormal swelling, and the urino-genital system in boys will be checked for undescended testes and foreskin problems.

The early years

Once the baby has left hospital, her development should be regularly monitored at the baby clinic. The clinic will assess the baby's physical development (sitting, crawling, walking, etc.), manipulative skills, sight, hearing, and language development. If there are any social problems, or if the parents have cause for concern, the child's social development will be assessed by observing the child at play.

The school-age child

When the child starts school, medical checks are carried out by the school doctor and nurse. A knowledge of the home background, and the teacher's relationship with the child, are often necessary to build up a complete picture.

The child is examined when she starts school. Her height, weight, sight, and hearing are tested, and a physical examination is carried out. Any child who needs follow-up checks will be looked at again in the future.

A child who is healthy will be examined again at the age of 13 to 14 years, but growth, sight, hearing, and teeth are regularly checked throughout schooling.

After the age of ten, or earlier in some cases, the child's health problems may include behavioural problems and drug-taking. In young teenage girls, problems may include pregnancy, anorexia nervosa, and so on. Any problems tend to be dealt with by teachers, school counsellors, or by the school doctor or nurse at the medical examination.

Once the child has left school, health screening is left more or less to the individual, and this is covered in chapter 21, 'Preventive health care'.

Assignments on health checks for children

1. There are certain conditions that may restrict a child's growth. Choose one of those listed below, and find out about it in more detail:
   - pyloric stenosis
   - cystic fibrosis
   - coeliac disease
   - diabetes
2. Some children fail to thrive for emotional reasons. Outline some home situations that may hinder a child's growth.
3. Babies are checked at birth to make sure that they are healthy. Sometimes a child may have phenylketonuria. Find out about this condition:
   - List the symptoms.
   - Explain the cause.
   - Outline the treatment.
4. A physical examination of a baby boy may detect undescended testes and foreskin problems. Find out about these conditions:
   - What are undescended testes, what is the treatment, and at what age is the child treated? What are the dangers of leaving the condition untreated?
   - What is a circumcision operation and when may it be medically necessary?
5. Sometimes it might be necessary to ask the doctor's or health visitor's advice about a baby's development. Very late development may be a sign that something is wrong.

   As a group, find out some of the causes of late development.
6. Health checks carried out on a pre-school child are intended to pick out any problems and treat them as quickly as possible. Look at these two areas and complete the charts:

   Movement/locomotion

Action	Approximate age
Sitting unaided Pulling up Cruising Walking unaided	

   What might be the possible causes of delayed development in any of these areas?

   Manipulative development

Action	Approximate age
Holding object Reaching out for object Transferring from hand to hand Pincer grip	

7. Screening the school-age child may pick up certain conditions. In pairs, find out more about the following conditions and report your findings back to the group:
   - colour blindness
   - scoliosis
   - enuresis
8. Many people feel that the needs of the teenager with problems are not well catered for. At present, teenagers may be helped by:
   - parents or teachers
   - school counsellors
   - school doctors or nurses

   With these limited options, it is quite possible that the teenager may find no source of help. What were your experiences at school? Could anyone who needed help have found somewhere to go?
9. As a group, find out about your local child health clinic. Include these questions:
   - When were child health clinics set up?
   - What are their aims?
   - What services do they offer?
   - What people make up the members of the clinic's council?

12 General illnesses in children

There are times when parents will need to ask their doctor's advice about their child's health. Parents, particularly the mother in most cases, know their child well and will be the first to realise that there is something wrong. Mothers can usually tell when their child is ill by changes in her behaviour.

Here are some of the symptoms that may be noticed:

- going off food or milk
- miserable behaviour
- lack of interest in surroundings
- sleeping more
- hot and feverish

These symptoms are usually a sign that the child is sickening for something, so the parents should keep a close eye on developments. More definite signs of illness are sickness, diarrhoea, and rashes. The parents will need to decide whether to contact the surgery. If the baby is young, it is best to contact the doctor whenever there is any concern, but as the child gets older the mother will be able to tell more accurately whether she is ill and needs to see the doctor.

Most doctors prefer patients to phone the surgery before 10 a.m., then the mother can be asked to bring the child in, or a home visit can be arranged.

Emergencies

In some cases, the child may need emergency medical help from the casualty department of the nearest hospital. Here are some conditions that will need emergency treatment:

- a blow, or heavy fall, to the head
- swallowing poisons, including medicines
- severe sickness or diarrhoea, especially in a baby
- swallowing something dangerous, such as a pin, etc.
- severe burns
- severe cuts
- fits and convulsions
- severe pain
- breathing difficulties
- choking

It is a good idea for parents to contact their own GP, where possible, before taking their child to hospital, so that any relevant medical information can be given to the hospital by the doctor.

If the child has swallowed poison or medicines, the container should be taken along to the hospital, so that the right treatment can be given.

Infectious illness

Although infectious diseases can affect people of all ages, there are some which affect mainly children. These diseases are carried by bacteria or viruses from person to person. In children, the diseases are less severe than in adults, and usually give lifelong immunity. In many cases the child can be given immunity to the disease by immunisation (see page 143).

Many of the diseases are spread by *droplet infection*, which is when small drops of moisture from the infected person are passed on by sneezing, coughing, or talking and may infect another person.

The so-called *contagious diseases* are passed on by direct contact, such as touching or kissing an infected person, or by indirect contact, such as using the same cup, or putting the same toys in the mouth, and so on.

Tetanus is an unusual disease, as it can only be caught by the germs passing into the body through a cut in the skin. That is why it is important that we all have regular tetanus booster injections.

Parasites

During childhood, children are likely to develop at least one parasitic condition. Perhaps the best-known of these is *head lice* (fig. 12.1), because the nurse regularly visits schools to inspect children's heads. Each louse feeds on blood from the scalp and leaves itchy bite-marks. The female louse lays eggs called *nits* which she firmly cements to the base of a hair. Lice are just as likely to affect children with clean hair and are not a sign of dirtiness.

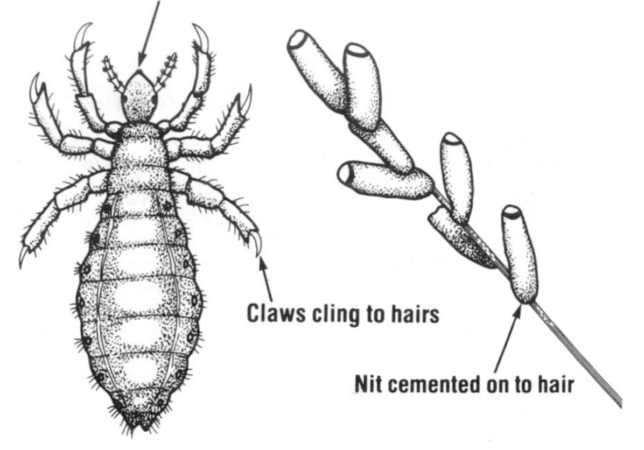

Fig. 12.1 Louse (about 3 mm long)

Human fleas (fig. 12.2) are less common than lice. They feed on human blood, leaving itchy bite-marks. Unlike lice, fleas like unclean surroundings and lay their eggs in unclean furniture, carpets, bedding, and clothing, etc. They will not live long in clean surrounding.

The worms that usually affect children are *threadworms* (fig. 12.3), which look like thin pieces of thread about 1 cm long. They cause itching around the anus when the female comes out at night to lay her eggs there. The eggs are able to survive for some time outside the body, and can be picked up by other people.

Scabies is a very irritating and itchy condition of the skin which is caused by mites that burrow into the skin to lay eggs. The rash, which appears three to four weeks after infection, is found in warm parts of the body such as between the fingers, on the wrists, and in the armpits and groins. The danger with scabies is that the severe itching may

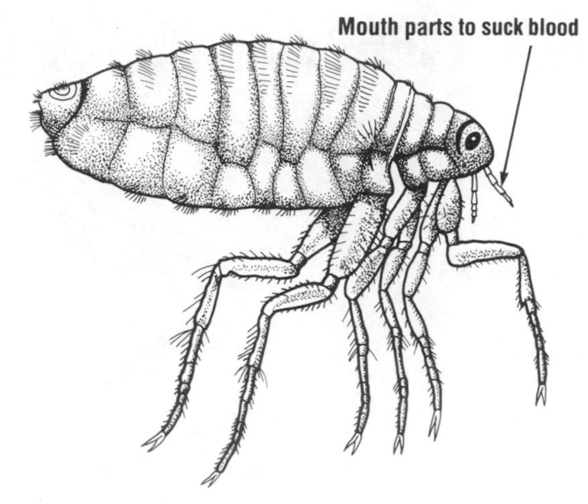

Fig. 12.2 Flea (about 3 mm long)

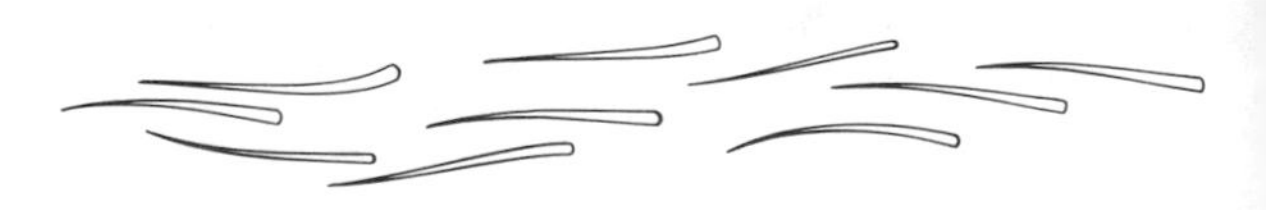

Fig. 12.3 Threadworms (about 10 mm long)

lead to secondary infections, such as boils and septic spots.

With all parasitic conditions, the whole family needs to be treated to prevent reinfection. Even after treatment the condition can return.

Allergies

General allergies, or the common allergic conditions of asthma and eczema, may affect a child's health. Allergic conditions may be present at birth or develop later.

If a person has an allergy, it means that he or she is particularly sensitive to the substance that causes the reaction, called the *allergen*. People may be allergic through their skin, through eating, or through breathing. Allergies, which often run in families, are caused by the body's being sensitive to certain, usually harmless, substances, which cause no reaction at all in people without the allergy. The body of an allergic person cannot differentiate between harmful and harmless substances. It may therefore make antibodies to fight these harmless substances, as well as to fight illness. A person with hay-fever will create antibodies against pollen, and will have an allergic reaction when pollen is present.

Caring for the sick child

Unless the doctor suggests that the child should stay in bed, there is no reason why she should not come downstairs where she will feel less cut off from life. She should, however, be kept warm and comfortable. Sometimes the child may prefer the comfort and quiet of bed.

Ill babies should be kept near the parents in a pram, if possible, as they will need a lot more attention.

Guidelines for nursing a sick child

Food Unless the doctor specifies what the child should or shouldn't eat, she should be allowed to eat as much or as little as she wants. Her appetite may be affected by the illness, and she may become faddy about food. Many mothers worry about food, and try to force the child to eat, but this really isn't necessary.

Drink Drinking *is* important, however, and it is essential that the child has plenty to drink. A baby may go back to a bottle during an illness, and this should be allowed. Offering a variety of drinks may encourage an unwilling child to drink a little more.

Room temperature The sick child should be kept in an even temperature that is not too hot or too cold. If the room is too hot, or if the child is wearing too many clothes, the heat may aggravate a temperature and cause convulsions. If the room temperature is too cold, then the child's body will have to work hard in order to keep her warm.

Clothing Clothing should be light rather than heavy, to ensure better temperature control.

Activities Time passes very slowly if the child is bored, and this usually happens when she is recovering. It is worthwhile for the parents to spend time thinking of things for the child to do.

Comfort The child should be kept clean and comfortable, with a routine that is as close to her normal everyday routine as possible.

Temperature Taking the temperature should only be used as a rough guide, as a child who is quite ill may only have a mild temperature, or none at all. Generally, a child's temperature is taken under the arm until she is old enough to keep the thermometer in her mouth.

Medicines Medicines are often prescribed by the doctor, and it is essential that the directions are followed carefully. It is important to think twice about giving the sick child medicines not prescribed by the doctor, and parents should check with the doctor or pharmacist first.

Hospital

Sometimes a child may be ill enough to have to go into hospital. The experience can be a shock, so it is important that she is as well prepared for the experience as time allows. An emergency admission allows no time to prepare the child, but a planned admission leaves plenty of time.

Apart from the fact that the child may feel ill, and may be in discomfort, the change in routine at the hospital is a great shock. Nowadays, efforts are made by hospitals and by various agencies to make a child's stay as pleasant as possible.

It is up to the parents to make sure – politely and tactfully – that the hospital is the right one for their child. Obviously, with an emergency admission this may not be possible. However, if the hospital falls short of their expectations, the parents can contact their local community health council, which will deal with their complaints and take any necessary action.

Once home, the child who has been in hospital may take a little time to settle down again. Until she comes to terms with the experience, she may revert to childish behaviour, such as bed-wetting, and be unwilling to lose sight of her mother. As with many problems, love and understanding will help the child to return to normal as quickly as possible.

Assignments on general illnesses in children

1. Children are quite frequently ill. Fortunately, the majority of childhood ailments are not dangerous. Find out about the symptoms and treatment of these common illnesses:

   - colds and coughs
   - sickness and diarrhoea (mild)

- conjunctivitis
- earache
- constipation

2. - As a group, check that you understand the meaning of the terms 'immunity' and 'antibodies'.
   - Discuss ways in which illnesses can be passed on by droplets, direct contact, and indirect contact.
   - How do we develop food poisoning?
   - What is the meaning of 'incubation period'?
3. Complete this chart on infectious diseases with details of chickenpox, measles, scarlet fever, German measles, mumps, polio, whooping cough, diphtheria:

Disease	Incubation period	Symptoms	Immunisation (if available)	Other comments

4. It is rare for very young babies to develop the infectious diseases listed in assignment 3, except for whooping cough.
   - Why are young babies immune?
   - Why are they not immune to whooping cough?
5. Parasites are very irritating, but not dangerous. Treatment is available for the whole family. Answer these questions:

   a) Head lice:
   - Draw a diagram of a head louse.
   - How do they breed?
   - What are the symptoms?
   - How is the condition treated?
   - How do lice spread from person to person?
   - How can we reduce the likelihood of catching lice?

   b) Fleas:
   - Draw a diagram of a flea.
   - How do they breed?
   - What are the symptoms?
   - How is the condition treated?
   - How are fleas spread from place to place?
   - Do the same facts apply to animal fleas?

   c) Worms:
   - How do threadworms breed?
   - How are they passed from place to place?
   - What are the symptoms?
   - How is the condition treated?
   - How can we reduce the likelihood of catching worms?

   d) Scabies:
   - Draw a diagram of a mite.
   - How do they breed?
   - What are the symptoms?
   - How is the condition treated?
   - How are mites spread from place to place?
   - How can we reduce the likelihood of catching mites?
6. As a group, check that you know the meaning of these terms:
   - allergies (give some examples)
   - allergic reactions
   - untreated allergies

   Find out about these common allergies:

   a) Asthma:
   - What are the signs and symptoms?
   - What are some of the possible causes?
   - What treatment is available?
   - What can be done to reduce the possibility of attacks?

   b) Eczema:
   - What are the signs and symptoms?
   - What are some of the possible causes?
   - Is there a cure?
   - What can be done to reduce the discomfort?

   What other substances may cause an allergic reaction in some people? What are the common reactions to these allergies?
7. The sick child at home needs to be kept occupied to avoid boredom. List some ideas that could prevent boredom for a five-year-old and for a ten-year-old.
8. Taking a temperature.

   a) The clinical thermometer:
   - Describe briefly how a thermometer

looks and draw a diagram.
- What is a normal temperature in °C and °F?
- Write a set of instructions for taking a child's temperature using a clinical thermometer.

b) The forehead thermometer:

- Describe this type of thermometer.
- Explain how it is used to take a temperature.

9. Answer these questions on giving medicines:

- Why is it important to follow the instructions on the medicine label?
- What should you do if the child shows some side effects from the medicine prescribed?
- Where should medicines be stored?
- Think of some ways to encourage children to take medicines.
- Why should antibiotics never be kept for use in the future?
- Why is it necessary to finish a course of antibiotic treatment?

10. The child being admitted to hospital for the first time faces some big changes. As a group, discuss the reasons why the experience may be a shock.
11. What questions should the parents ask about the hospital their child is being admitted to? Here are some possible ideas:

- visiting
- 'rooming in' for mother
- play and education
- food
- clothing

12. Ideally, children should be prepared for hospital well before they ever have to be admitted. As a group, suggest ways in which hospitals can be introduced to the very young child.
13. Sometimes it is necessary to prevent infectious illnesses from being passed on to other people. Put together a set of guidelines to help prevent the spread of infection from a sick child to others.

13 Adolescence

Adolescence is the name given to the span of time from the start of puberty to adulthood, which is usually the ages of 10 to 18 years.

Puberty is the term used when a person becomes physically able to reproduce. For girls, puberty usually starts at around the ages of 10 to 13, and for boys at 12 to 14 (fig. 13.1). In some cases, puberty may start earlier or later, but generally this is nothing to worry about.

The age when puberty starts is determined by many factors, including heredity and diet. Nowadays, boys and girls reach puberty a couple of years earlier than was usual at the turn of the century, and part of the reason is thought to be a better diet.

Before discussing puberty, it is important to understand the male and female reproductive systems, and these are discussed in chapter 2.

Puberty is a time of increasing sexual awareness. When a girl starts her periods and a boy has an ejaculation, they are said to be sexually mature, as they are able to reproduce. That is setting sex aside from emotional maturity: there are very few boys and girls emotionally ready for parenthood, or even a deep relationship, at this stage.

During the teenage years, many emotional

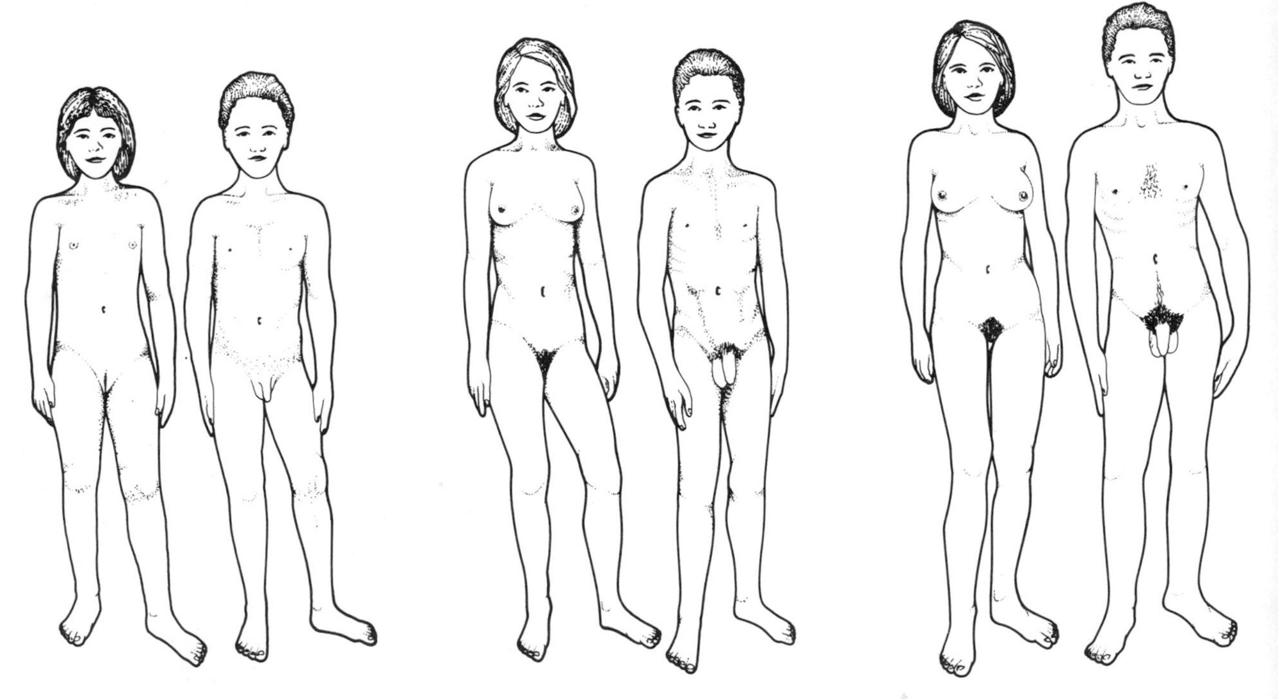

Fig. 13.1 Boys and girls at puberty

changes take place. Many teenagers fall in love for the first time, and find that every waking minute is spent thinking about the other person. This is called 'infatuation', and it can be a very exciting and happy time, especially if the other person returns the feelings. This type of love is often blind: the teenager sees only the good points until the time when the other person's true character becomes apparent – then the infatuation may disappear. The feelings left are often irritation, dislike, and disbelief that he or she could ever have seen anything attractive in the other.

Many adolescents pass through a homosexual phase, where they form strong and emotional relationships with people of their own sex. This phase is part of a growing and developing awareness of a need to give and feel affection, and not necessarily a sign of true homosexuality.

During adolescence, most teenagers go on to form heterosexual relationships. At this stage, it is necessary to bring the term 'responsibility' into the discussion, as forming a relationship means that the feelings and needs of two people are involved. The couple need to understand each other's differing sexual needs. The boy may be more easily sexually aroused than the girl, but it is not fair to use this as a means of persuading her into intercourse. The girl may see intercourse as one part of love-making – not the only part – and may find it difficult to understand the boy's urgency. It is important that the couple consider each other's feelings at this stage in love-making, and in all other areas of their relationship.

Love itself seems very strange to adolescents. They see their own relationships develop and fade, and see parents and friends of the family getting divorced. Many teenage girls have false expectations of love, based on the books and magazines they have read: they may see marriage as the be-all and end-all of a relationship, and believe that love is enough to change any man. Sadly, this is not so: all relationships involve work and commitment if they are to succeed. Each partner has to make an effort to understand the strengths and weaknesses of the other, while trying to make himself or herself a better person to be with. There will be problems and even crises in a relationship, and these may cause some partnerships to break up. However, if a couple work together to overcome their problems, their relationship will often strengthen and develop. Sex itself is only part of a strong relationship: it is not enough to build a relationship on, but a very important part of a loving, stable relationship.

Peer pressure (the type of pressure exerted by people the same age) is strong, as nobody wants to be different. Groups of teenagers look for an identity, which may be based on a pop culture or current trend. Friends are made with people of similar tastes and outlook. As they grow older and reach their twenties, most teenagers are less dependent on a group identity and have formed their own independent opinions, views, and beliefs.

These years are the time many teenagers have problems with their parents, as young people want to assert themselves and reject the values of their parents. They also realise that adults are not always right – they make many mistakes and are not always secure and happy. Parents, on the other hand, fear that their children will repeat the mistakes they made and try, fruitlessly, to stop their children finding out the hard way. Some parents may feel jealous of their child's freedom and lack of responsibilities, and long for their youth again so that they can avoid the mistakes they made in their past.

It is important that both the parents and the teenager should realise the difficulties and try and understand the other's views in order to help to resolve the situation. The majority of parents are only trying to shield their child from an unpleasant experience.

Problems

Adolescence is a time when the teenager may have to cope with many problems for the first time.

Skin problems, such as spots and blackheads, are caused by the hormonal changes at puberty. Although they do disappear in time, that is no comfort while they are there. Acne, which affects about 90% of teenagers, may be on the face, back, and chest. Acne can be severe, leading to pitting and scarring. Although many of the creams, washes, and ointments for sale in the chemists do help some people, unfortunately there is no real cure. However, spots and blackheads are more noticeable to the individual concerned than to other people.

Weight problems cause much anxiety, especially for girls. Slim models with perfect bodies appear in all the advertisements and magazines, and make many teenage girls feel fat and ugly, or skinny and unattractive. Many adolescent girls (and this rarely happens to boys) become obsessed with slimming, and lose touch with what is really happening to their bodies. This condition is called *anorexia nervosa*. The girl either refuses to eat or, if she does, she will make herself sick in order to remove the food, or she will take large doses of laxatives. Another situation that arises is one where the girl might eat vast quantities of food and then bring it up in a fit of remorse. This is called *bulimia*.

If the situation continues, the girl's periods will stop and she will get thinner and thinner – yet see herself as fat in the mirror. Many doctors believe that the anorexic has an underlying psychological fear of becoming a mature woman and so diets drastically in order to keep a child's weight and stop her periods.

Aggression and anti-social behaviour

Adolescence is often a time of anti-social behaviour, where the teenager may show signs of disobedience and aggression. In most cases, this is symptomatic of achieving independence, and is a temporary phase. In fewer cases, if the behaviour becomes persistent, it may be the sign of a problem. The teenager will need to get help from a social worker, psychiatrist, or doctor, etc. Social problems, such as drug and alcohol abuse, are discussed in chapter 18.

Hygiene

During adolescence, the body alters because of hormonal changes preparing it for adulthood. There is a greater need for what advertisers delicately call 'intimate hygiene'.

Everyone needs to keep himself or herself clean. By the time most boys and girls reach puberty, they have passed the phase where they need to be reminded to wash – although there are always some who are happy to be dirty. For the majority who want to look and smell nice, there are some facts that need to be understood.

Within the vagina, natural secretions keep everything clean, so there is no need to wash there. Once the secretions reach air, they are broken down by bacteria, and may smell unless the area is washed regularly with mild soap and water.

Men need to keep themselves clean for two reasons: firstly because it is more hygienic, and secondly because they may transmit infections such as thrush and other bacterial infections to their partners during sexual intercourse. It is particularly important that uncircumcised men should keep the area behind the foreskin clean, where secretions collect to form *smegma*. Research has suggested that lack of hygiene in this area may be a contributory cause of cervical cancer.

Sexual relationships

Because many adolescents may be engaged in sexual relationships nowadays, there are an increasing number of teenage sexual problems. Sexually transmitted diseases (STDs), which are discussed later, in chapter 19, are usually associated with promiscuity, yet they can result from one brief encounter with an infected partner.

Schoolgirl pregnancies are increasing, with 5000 under-16s becoming pregnant each year. Of these pregnancies, about two-thirds are terminated by abortions. This is a very traumatic experience for a young girl and, whether she decides to keep the baby or have an abortion, she will need help, advice, and counselling.

Masturbation

During adolescence, boys and girls discover their developing sexual awareness. Research states that most boys and girls *masturbate*, which means bring about their own sexual pleasure.

Although it is thought that more boys than girls masturbate, this could have something to do with the fact that girls tend to talk about sexual matters less than boys. Masturbation is often called 'self-abuse', which gives a rather critical and moralising impression. Medically, it is described as a means of creating sexual arousal, and sometimes orgasm, by handling one's own genitals.

However, for most children, masturbation starts well before puberty, as many young children find pleasure in handling their genitals. It is at this stage that they may also learn to feel guilty, as some parents will tell them to stop as it is 'rude and dirty'.

At puberty, this interest in the sex organs returns because of the major changes in the body, and changes in the hormone balances.

The question remains whether masturbation is right or wrong, good or bad. In the past, it has always been frowned on, and a feeling of disapproval still exists to some extent today, even though many adults masturbate despite a stable sexual relationship. Many sex experts believe that masturbation is valuable as it helps boys and girls to understand and handle their own bodies, which will help them to form meaningful sexual relationships in later life. However, others believe that there is a danger that too much masturbation, being a solitary act, may make it more difficult for the person to relate to another adult in a relationship. It is important that we reach a balanced understanding of the situation in order to avoid feelings of guilt. Masturbation in adolescence is a perfectly natural and normal part of growing up and understanding one's own sexuality. Masturbation in later life, although not causing any physical damage, can be a sign that there are underlying problems in a couple's sex life, especially if it is done in preference to a joint sexual act.

Erotic dreams

Adolescence also sees the beginning of erotic dreams, which are sexual dream fantasies. In boys, these are called *wet dreams*, or *nocturnal emissions*, if they are accompanied by an ejaculation. In girls, they are less noticeable as there are no external signs. It is rare to be conscious of these dreams, and most of the time adolescents are unaware that they have had one.

Adolescence is a time of growing and changing, which can lead to problems. However, there is much that the adolescent and the parents can do to help make the changes as smooth as possible. Good communication, understanding, sympathy, and tolerance are needed on both sides.

Assignments on adolescence

1. Although there is more preparation for parenthood in our schools than previously, not enough is taught about the roles and responsibilities of forming a mature sexual relationship. In pairs, plan a useful educational programme aimed at preparing adolescents for:
   - relationships
   - needs of children
   - practical child-care
   - family life

   You need only plan the programme – do not go into detail. Compare your headings with those of the other pairs in the group.
2. In certain countries, an operation called 'female circumcision' is carried out on baby girls, for religious and cultural reasons. This operation is not quite like the male circumcision, as the entire clitoris is removed – thus denying the female any sexual satisfaction. As a group, find out more about this operation, and discuss the moral implications.
3. 'Virginity' is a much-debated issue. Some men still think it is important to marry a virgin, while believing that a man should 'sow his wild oats'. Discuss this apparent contradiction.
4. In small groups, discuss your first love. Remember the feelings and emotions you experienced. How do you see the situation now?
5. As a group, list the various youth cultures that have existed in the past 30 years (e.g. teddy boys, punks, etc.):
   - What was their common interest?
   - What clothes were associated with each movement?
   - What music was associated with each movement?
6. What problems can arise between parents and their teenager? In small groups, discuss the problems and arrange them under different headings. Discuss ways in which parents and teenagers could work through these problems to create a better relationship.
7. 'Responsibility' is a much-used word nowadays. Discuss the term when applied to a young couple's relationship. How should it affect the way the boy and girl see each other?
8. Obtain some copies of romantic books and magazines. In small groups, take one of the copies and analyse the type of stories in it. Look for any of the following, and add to them if you wish:
   - Is there a happy ending?
   - What type of descriptions are used for the boy and girl or man and woman?
   - Are these descriptions stereotyped, and, if so, how?
   - Is the situation true to life?
   - Do you see love like this?
9. Write a set of 'dos' and don'ts' for the control of acne.
10. Anorexia and bulimia nervosa are conditions that affect more girls than boys. In small

groups, discuss the conditions:

- Have you ever experienced the need to diet?
- Do you know anyone who has had anorexia nervosa or bulimia nervosa?
- How should a parent cope with a daughter who has anorexia nervosa or bulimia nervosa?

11. Douches and vaginal deodorants are bad for women and unnecessary. Find out what they are, and the dangers of using them.
12. The Victorians believed that masturbation reduced stamina. Christians and Jews believe that the Bible condemned it by the story of Onan in the book of Genesis. Some parents tell their children that masturbation makes you blind or deaf. As a group, discuss your opinions of such stories. Do you think they may be harmful, or might there be some truth in them? What would you tell your children?
13. Schoolgirl pregnancies are increasing in number every year. Girls who decide to keep their babies need plenty of support. Give advice to a girl on how to cope with her pregnancy, under the headings of financial advice, emotional advice, medical advice. Include ante-natal care, financial help and accommodation, the girl's education, and the future of the baby.
14. Many women say that men get away pretty easily when things go wrong in a relationship – after all, it is the woman who is, literally, left 'holding the baby'. As a group, discuss your views on this. Include also the position of the schoolboy father of the baby in question 13.

14 Adulthood

Leaving home

After the search for an identity which takes place during adolescence, the majority of young adults find themselves concerned with getting on in life in order to gain financial and emotional independence. Most will either have left home or will be thinking about it. Leaving home is quite a traumatic experience, as it is the first major step into a new life. In making a new home, the young adult will either try to make a replica of his or her family home or will break free entirely and create something very different. Either way, most young adults will feel homesick at times, mainly through feelings of insecurity and fear of the new.

In time, if they do not give in to these fears, young adults will come to terms with the situation. They will accept the new home as 'home', and be glad to be there.

Parental attitudes affect the situation: parents who genuinely want their child to gain independence will offer support and help where necessary, and will not try to prevent their son or daughter from leaving home. In fact, if parents can see this period as an opportunity to spend more time with each other rather than the loss of their child, then they will have a far healthier attitude to what is, after all, a natural progression from adolescence to adulthood.

Forming a steady relationship

By the time they are in their twenties, most people have formed a steady relationship, whether heterosexual or homosexual. In the heterosexual relationship, the couple have the option of living together, getting engaged, or getting married. Although living together is quite acceptable in our society, many couples still decide to get engaged, even if they are having a sexual relationship, as they enjoy the security the engagement gives them. Eventually, the couple will probably get married.

People marry for many reasons, and love is only one of them. Here are some of the reasons that people may marry for:

- Sadly, many people, especially girls, marry to escape an unhappy home background. They feel that the only way to find happiness is through marriage, although the reality is often that they have swapped one set of restrictions for another.
- Some people marry for companionship, as they find living alone intolerable. They hope that the right marriage partner will be able to offer all the love, security, and strength that they need, and they are often bitterly disappointed when they find that their partner is unable to live up to these expectations.
- Many young marriages take place because friends are getting married, and it seems the fashionable thing to do. It is very exciting to plan a wedding and choose the various outfits, and very easy to be bitten by the marriage bug. Again, such an unrealistic view of marriage is more likely to be disappointed in time.
- Even in today's more enlightened times, a few people – and, again, especially women – feel that they are not socially acceptable unless they are married. There is still the old idea that a single woman has been 'left on the shelf' and nobody wants her. In a way, society is very much to blame for this attitude, as families are often classified according to the status of the father/husband: the wife's status reflects that of her husband.

Men are not completely free of the stigma of being single. Many firms prefer to employ a man who is married once he has reached his mid-twenties, as the feeling is that he will be more steady and reliable.

Homosexuality

In Great Britain, homosexuality is legal between

consenting male adults of over 21. There is no law on female homosexuality (lesbianism) as, at the time of making the laws, Queen Victoria did not accept that women could be homosexual.

Homosexuals usually encounter a great deal of prejudice from society, as there is a belief that children are at risk, but nowadays many more homosexuals are 'coming out' and being open about their situation.

Although society has become more tolerant towards homosexuality, many people still fear homosexuals and brand them as perverts. However, it is important to remember that people differ sexually, in the same way as they differ physically. Both sexes have male and female hormones and hence both 'male' and 'female' aspects to some extent; but, in general, a man has more male hormones and a woman has more female hormones. During adolescence most people develop a preference for the opposite sex, and form *heterosexual* relationships.

Even between homosexuals, there is a great deal of variation. Some may be interested only in people of their own sex, whereas others may be interested in either sex and are called *bi-sexuals*. Quite often a married person may have an affair with someone of the same sex while remaining happily married. A heterosexual person may develop homosexual tendencies if put in a situation where the only sexual opportunities offered are homosexual – in prison or at sea, for instance.

Research has come up with many theories to explain homosexuality, including over-dominant parents and brain malformations, but it is far more likely that homosexuality is part of an individual's genetic make-up rather than a result of his or her upbringing.

The secure relationship

There is no way of ensuring a successful marriage, or relationship. However, statistics show that young marriages are more likely to fail. They also show that couples with a similar background are more likely to have a happy relationship, but statistics do not take into account human differences.

Individuals are growing and changing all the time throughout their lives. Various experiences affect and alter our views, which will obviously affect the partner in a relationship. A good relationship will be able to withstand these changes, and will grow stronger as each person learns more about the other.

Coping with these changes will probably involve arguments, but anger and frustration are healthy emotions as long as they are talked about and not bottled up. In a good relationship, the couple should be able to communicate their anger and put it to positive use.

After a few years, the initial romantic and exciting type of love will gradually diminish and, ideally, the couple will have already found other, equally satisfying, aspects of their relationship. Sadly, when they reach this stage – often referred to as the 'seven-year itch' – some couples may look for excitement in affairs. Other couples reaching this stage in their relationship find that newer interests, such as a family, a new house or job, and so on, will help to satisfy their need for excitement.

Couples with problems who feel they would still like to make a go of their relationship often find marriage-guidance counselling helpful. Outside help can be very effective in pointing out facts that the couple themselves are unaware of.

Divorce and separation

The break-up of a relationship is a very painful experience, no matter how bad the relationship was. In our present legal system, a divorce can become a battleground, with each partner trying to get the bigger share. In some areas, attempts are being made to create conciliation services, which aim to help the couple part as amicably as possible. These services cost money, so are not freely available throughout the country.

At present, one in three marriages in this country ends in divorce, and these figures do not take into account the number of cohabiting couples who split up. This figure means that almost two million children live in one-parent families.

Divorce is a time of great anguish, and is second only to bereavement on the stress scale. It may cause mental and physical ill health in some people. For these reasons – consideration of any children and personal health – conciliation may become a necessity in the future.

The single person

Although most people will marry and perhaps have a family, a growing number choose to remain unmarried.

In the past, to be unmarried was considered

unusual – the person was labelled as being 'left on the shelf' by the opposite sex. This was particularly so for single women – the term 'spinster' still has unpleasant connotations even today. Unmarried men were more fortunate, as the term 'bachelor' was associated with freedom, fun, and escape from marriage.

Over the past 20 to 25 years, women have become more financially independent and no longer need to find a husband to support them. It is no longer considered unusual, or immoral, for a single woman to have a baby. These changes in society's attitudes mean that it is now quite acceptable to remain single – there is no longer the pressure to conform.

The childless couple

In the past, a married couple were expected to have a family soon after marriage. Couples who didn't have a family were often considered to be either selfish or materialistic. Most women did not expect to have a career but saw their future security in terms of marriage. Those who did work usually gave up their jobs as soon as they got married. Contraception was used to space babies rather than to prevent them altogether.

In the 1960s, various changes occurred that altered the pattern of family life:

- Women became more aware of their rights, and many decided to pursue a career after marriage. In the 1970s, maternity leave made it possible for women to return to work after the birth of a baby.
- More effective contraception was freely available, so couples could remain childless.
- Moral standards altered, and it became acceptable for couples to live together, rather than marry.
- It became easier to obtain an abortion.
- Women received equal pay for equal work, so it became easier for a woman to be financially independent.

The result of these changes is that it is no longer considered unusual for a couple to want to be childless – the decision is theirs.

Starting a family

When a couple marry, or form a steady relationship, friends and family tend to expect them to have a family. Even in these days of efficient contraception, many people will start to ask questions if the couple are still childless after three years or so. A couple who have decided to have no children will often be thought of as selfish or 'unnatural' – our society likes people to 'follow the norm'.

Choosing when to have a family is ideal, but many babies are conceived with little or no thought on the part of the parents. In a way this is a pity, as, if it is what they both want, planning a baby can bring a couple closer. However, it is not always possible to become pregnant to order (see the section on infertility in chapter 2) and many couples have to come to terms with the fact that they will never be able to have a family.

Unfortunately, some people have a baby for the wrong reasons:

- Some people hold the mistaken view that a baby will bring together a shaky marriage. This usually isn't the case: the strain of caring for a baby can often test the most secure relationship, and may well ruin a less happy one.
- Some people believe that having a baby proves their sexuality. A man may feel he has to prove his virility by fathering a child, and a woman may think that the only way to demonstrate her femininity is by having a baby.
- A small number of parents will have a baby to create some sort of excitement in their lives. They enjoy the attention they get during the pregnancy and just after the baby is born, but they may find that the excitement is short-lived.
- Another category is the couple who have a baby because it is expected of them, perhaps to please the grandparents.

Once the baby is born, many couples feel a sense of anti-climax after the weeks of waiting. They have to adjust to their new life together as a family. The woman may be suffering from baby blues or, worse, post-natal depression (see page 44). The man may feel jealous of the attention the new baby is getting. However; these problems are usually temporary, and a stable relationship will survive them.

Early middle age

As people reach their thirties, they are aware that middle age is around the corner and that they are no longer young. They become aware that many of the opportunities open to the young are no longer open to them. High-powered job advertisements often ask for someone aged 25–30. For many people, if they haven't reached their mark by their mid-thirties, they may be passed over in favour of someone younger.

The years between 30 and 39 are a time for people to take stock of their lives and assess how successful they have been in terms of careers, money, relationships, and happiness. Couples who have decided to wait before starting a family need to act soon. Single people wonder whether they have made the right decision, as do their married counterparts. If there are any life changes to be made, there is the feeling that they need to be made now, before it is too late.

Middle age

Middle age can be a very strange period of a person's life. As parents, the middle-aged may have teenage children to care for and, as children themselves, they often have the responsibility of their own ageing parents. Parents see their own children becoming young adults, and their ageing parents remind them of their own fate in a few years' time. It is the time of life when men and women often wonder just what they have achieved in life, and these thoughts may lead to feelings of guilt and depression. This time is often referred to as the 'mid-life crisis', a title which reflects the depth of feeling about middle age.

Many middle-aged parents, particularly mothers, feel that they are now useless. This belief may be caused by the fact that they have lived their lives too much for other people and have not considered themselves enough. A more positive attitude towards middle age is to see it as a chance to do the things that there was no opportunity to do before.

A negative response to the mid-life crisis is to try to recapture lost youth, rather then accepting and enjoying the middle years. Adults may feel certain regrets: perhaps they were not as successful at work as they had hoped, or possibly they may feel bored with their marriages, and long to recapture some of the romance they see around them. Whatever the cause of these feelings, the result is a sense of dissatisfaction which can have a detrimental effect on the person's life. Older men may find themselves turning to younger women in an attempt to prove to themselves that they are still attractive to the opposite sex. Women may find themselves wanting to completely change their appearance. Although these examples may seem to be stereotyping men and women's reactions to the mid-life crisis, there is more than a little truth in them.

In order to cope with these destructive feelings, the adults concerned need to be able to accept themselves for what they are and to be reasonably satisfied with what they have achieved in their lives, rather than worrying about what they haven't done.

In the same way that periods marked the start of their reproductive life, women go through a menopause, or 'change of life', to mark its end. There are a lot of myths surrounding the menopause which cause unnecessary fears and worries for many women. It is not true to think that a woman cannot enjoy her sex life during or after her menopause. In fact, many women say that their sex life improves, as there is no longer the fear of becoming pregnant. Any problems that do occur can usually be helped by a visit to the doctor, or by an understanding partner.

For both men and women, a decrease in sexual activity is more likely to be due to other preoccupations, such as work, health problems, or too much smoking and alcohol, rather than to middle age itself.

Between the ages of forty-five and sixty, the otherwise healthy adult will, by the law of averages, be more likely to suffer from ill health. The death rate doubles, largely due to cancer, heart disease, strokes, and respiratory and circulatory disorders. However, the chances of developing these illnesses can be reduced by eating a healthy diet and taking regular exercise. Even if these changes in lifestyle do not take place until middle age, they will help to prevent any existing conditions from getting any worse.

Assignments on adulthood

1. Through role-play, act out the following situations in which the young adult leaves home:
   - over-protective parents, unwilling to let go of their child: mother sees this time as the loss of her 'baby', and father tends to agree
   - supportive parents who realise that leaving home is part of growing up: they are willing to help in any way they can
   - uncaring parents who lead their own lives:

child feels forced out of the family home

Analyse your views on each of these three situations. Use your findings to make a set of guidelines which will enable parents and young adults to cope with leaving home.

2. As a group, discuss your views on young marriages.
   - Do you think the couple should live together first?
   - Do you think that couples should wait until they are in their mid-twenties before marrying, as is the case in China?
   - Statistically, young marriages are more likely to end in divorce. Why is this?
3. People marry for many different reasons:
   - Can you think of any other reasons for marrying apart from those covered in the main text?
   - Of the four reasons covered – companionship, escape, fashion, and social acceptability – pick one, and discuss it in small groups. Include in your discussion:
     - a few examples of circumstances that may lead the young adult to marry for this particular reason
     - why this may not be the basis of a successful marriage
     - ways in which you could avoid making similar mistakes
4. In some societies, the marriage is arranged by the parents when the bride and groom may still be children. Although these marriages are not based on love, they are usually considered to be successful. Discuss your views on arranged marriages.
5. Discuss the ways in which a relationship may be damaged by bottled-up anger and frustration.
6. What advice would you give to these people and their partners:
   - a jealous and possessive husband, who refuses to let his wife go to work as he believes her place is at home, where she is away from the outside world, and temptation
   - an insecure wife who worries that her husband is having an affair
   - a husband who beats his wife
7. One in three marriages ends in divorce. In order to make a divorce as painless as possible to all the parties concerned, conciliation counselling may be helpful.
   - Find out what you can about conciliation services.
   - Are there services available in your area?
   - What help is available in your area for a couple going through a divorce?
8. There are many reasons for divorce. Here are a few of them:
   - a partner's changing opinions or attitudes
   - a new family
   - extra-marital affairs
   - incompatibility
   - too high expectations of marriage
   - financial problems
   - marrying too young

   For each of these factors, think of any ways a couple in your area could be helped. Sources of help could be:
   - outside help, e.g. counselling, conciliation, etc.
   - friends and/or family
   - legal help
   - benefits
   - national or local support groups
9. The only ground for divorce today is 'irretrievable breakdown', which is supported in five ways. What are the five ways?
10. How soon after their wedding can a couple start divorce proceedings?
11. Find out about these terms: petitioner, respondent, decree nisi, decree absolute, affidavit, child custody, maintenance.
12. What do you think are the right and wrong reasons for having a baby?

    When you have discussed this, write a list of what you believe parenthood should involve.
13. What are the physical and emotional signs of the female menopause? Discuss ways in which women can prepare themselves for the menopause so that it causes them as few problems as possible.
14. Although men may not have the equivalent to the female menopause, what physical and mental problems may a man have to cope with

in middle age?

15. Recently, there have been debates in medical circles about the value of hormone replacement therapy (HRT). Find out what you can about HRT and, as a group, discuss your views on its benefits.
16. As a group, devise a questionnaire to find out about middle age. It would be useful to separate your results into early and late middle age, to compare differences. Decide on the areas you want to research, e.g. lifestyles, career, attitudes, politics, health, etc., and formulate your questions so that they can be answered, as far as possible, with a 'yes' or 'no'.

15 The ageing process

The process of ageing is very gradual. Every minute we lose nerve and body cells that cannot be replaced – in other words, we start to age the minute we are born.

However, in this chapter, when we talk about 'ageing' and 'old age', we refer to the time when the ageing process has become noticeable and apparent.

Outward appearance

Usually, the skin of the older person will have lost much of its elasticity, which causes wrinkles and lines to appear all over the body. The most obvious of these are on the face, especially around the eyes and mouth, and are often called 'crows' feet'. In the much older person, the skin will tend to be drier, and there may be brown patches of pigment.

Perhaps one of the first signs of the ageing process is that the hair may lose its colour and turn grey. This is caused by the gradual loss of the pigment which is found in each strand of hair. It is not uncommon to find people in their twenties or thirties whose hair is grey.

Other, less immediately noticeable signs are that the elderly will tend to bruise more easily, and will sweat less, than a younger person.

The body

As you look around at old people, and study the old people who are part of your family, it is clear that they no longer have the agility or strength that they had when they were younger. This is because the muscles very gradually shrink and waste away as we grow older.

The bones of the elderly may become very brittle and are more likely to fracture. This is caused by a reduction in the amount of calcium found in the bones, a mineral essential for the strength and durability of bones.

These changes in the bones, together with the muscle wastage, cause old people to lose height and literally shrink. Old people tend to be stooped and bent, and many are shorter than they were in their youth.

More serious problems of the bones and joints can lead to arthritis, which will be covered further on in this chapter, under 'Medical problems'.

Another common feature of old age is loss of memory, particularly what is known as 'short-term' memory. It is this that causes so many old people to be forgetful about the present, but to be able to recall their childhood, or wedding day, as if it were yesterday. When this memory loss becomes worse, the old person may be suffering from 'confusion', dealt with later under 'Mental-health problems'.

Like all parts of the body, the brain is composed of millions of cells, all of which help the body to function efficiently. As the person grows older, the number of brain cells decreases, and they are not replaced. The more immediate results of this are that the muscle co-ordination and balance of old people may be affected, and this slows them down. They will also take longer to adjust to their immediate environment, such as extreme heat or cold, which could result in hypothermia (see later in this section) or they may burn themselves if they sit too near a fire. Because their reaction time is slower, it is essential that extra safety precautions are taken in order to reduce the risk of an accident.

The senses, too, are affected by the process of ageing. The sense of touch becomes less accurate, so the old person is less likely to be aware of pain. Again, this could lead to burning etc. while the old person is unaware of the damage. Eyesight frequently deteriorates, as we can see by the number of elderly people wearing glasses. Hearing problems become more common as we grow old, although hearing-aids can alleviate this handicap.

Internal organs

The heart, which is the muscle used to pump blood

around the body (see chapter 1), becomes less efficient in old age. This means that less blood is pumped around with each beat of the heart, so less blood reaches the body tissues. The outward appearance of this may be the mottled skin usually associated with a 'bad circulation'. A poor heart can lead to greater problems, discussed later in this section.

The lungs, which are used to take in air from the atmosphere, also alter with age and become less efficient. The old person will only be able to manage a shorter intake of air, and as a result will become breathless more easily.

As well as taking in less air with each breath, the lungs also work less efficiently. Once the air has been breathed in, the lungs remove oxygen to keep the body functioning well. However, in the elderly, less oxygen is removed from the air, so old people find it more difficult to do anything energetic. Inefficient lungs can lead to a greater, more dangerous, problem – the heart has to work much harder to make up for the lack of oxygen from the lungs, and this puts a strain on it. The results of this are dealt with later, in the section on 'Medical problems'.

The digestive system, although it is inside the body, is mainly made up of muscles. We have no control over these muscles as they work under orders directly from the brain, unlike the muscles of our arms, legs, and so on. Earlier, we read that the muscles change with the ageing process, and this is also true of the muscles of the digestive system. The muscles become less efficient, which causes two main problems:

1. The food in the stomach is not reduced as much as it should be by the action of the stomach-wall muscles. The result of this is that much of the food remains undigested, so the old person may lose weight.
2. The muscles of the gut wall lose their strength, so it becomes harder for them to push the food along the alimentary canal. This will cause the food to stay in the gut for longer than it should, possibly resulting in constipation.

As the heart becomes less efficient at pumping blood around the body, the rate at which blood flows through the kidneys is lower, so the rate at which waste products are removed from the bloodstream is lower. These waste products will therefore build up in the bloodstream and will affect the healthy functioning of the body cells. (It is thought that the kidneys of a person of 80 are only half as efficient as those of a 25-year-old.)

The metabolism of the elderly person slows down, due to various changes in the glands. (Metabolism is the rate at which the body burns up food and oxygen, and is dealt with on page 8.) When the metabolism changes, the old person may feel tired and lethargic and lack the energy needed to do everyday tasks. This need not cause problems, as long as the person comes to terms with his or her limitations and does not try to do too much.

Perhaps one of the more publicised problems of old age is *hypothermia*. As has been said before, in the elderly, the body becomes less efficient at maintaining a constant temperature. In summer, an old person may quite happily wear a thick winter coat. In winter, however, there are more dangers as the old person loses body heat. If the person's temperature drops too low as a result of the cold, it may not rise again without help – this condition is known as hypothermia.

Throughout this century, scientists have been particularly interested in researching into the ageing process. They have been studying the physical causes of ageing, and have so far tended to concentrate on the basic unit of all living things, the cell. Many scientists now believe that, as each cell reproduces itself, it becomes less perfect each time, resulting in ageing of the body tissues.

Another theory suggests that the protein present in the body cells may alter and create less efficient cells – again resulting in the body's ageing.

The immune system in our bodies, which fights off infection and disease by producing antibodies, could become directed towards its own cells – a possible cause for tissue changes in old age.

It is likely that there is no one cause of ageing, but that it is a combination of these three theories plus many other factors. After all, why is it that some people look young for their age, whereas others seem to age early? The way we care for our bodies affects the way we look, and therefore must have some bearing on the speed at which our bodies age.

Assignments on ageing: the body

1. In class, discuss the following terms to check that they are fully understood, as some of them

will be referred to again in the following sections:

body cells	pain threshold
proteins	heart and circulation
immune system	digestive system
antibodies	urea
pigment	metabolism
muscles	hypothermia
bones and calcium	heredity

2. In small groups, use the various causes of the ageing process as the basis of a discussion. You could include some of the following points in addition to your own:

- Why do scientists research into old age?
- Is there anything that can be done to slow down the ageing process?
- Do you think heredity affects the speed at which we age? Give some examples.
- Do you think scientists will ever find the secret of eternal youth? What do you think about all the cosmetic creams on the market that promise to make your skin look younger: can there be any truth in their claims?
- Do you think plastic surgery, such as face-lifts and removing bags under the eyes, etc., is a good thing? Should we be able to have these operations on the National Health Service?

Another side of old age

We have looked at the external and internal effects of ageing on the body, but what about the other aspects of old age, such as feelings and attitudes, and social issues? In books about old age, these aspects are frequently overlooked, yet to the individual concerned they are very important.

In most Western countries, old people are forced into retirement at a certain age, regardless of their physical or mental health. This may cause them to feel useless, and to be unsure of how to pass all the extra time they have on their hands. Also, in our civilisation, only a minority of old people live with their families. In the past (and in third-world countries today), the elderly were respected and valued as people and were included in the everyday life of their families. They still had a valuable role to play – which is an important factor in preserving mental health; they could keep their self-respect. Nowadays, the elderly all too often suffer from some form of mental ill health.

The mind, like the body, undergoes changes as part of the ageing process, and these changes vary according to the individual. Because the brain slows down, old people tend to take longer to cope with problems. As it takes longer to learn something new, they prefer to dwell on what is already familiar to them. As a result of these changes, they may become stubborn and intolerant of anything new, which are two of the personality changes frequently associated with the old.

It is all too easy for us to say, 'This can't possibly happen to me', but, unless we take positive steps, it probably will. Education for old age needs to start in our youth. It is a good idea to develop a wide range of interests so that, when retirement does eventually come, the older person is not left with a vacuum to fill. It is also important to keep the mind stimulated, as stimulation is the food of the mind and brain. Again, preparation for this starts well before old age comes: if we were to educate ourselves earlier for old age, many of the problems associated with it would decrease dramatically.

Loneliness is another problem common in old age. As time passes, everyone has to face the fact that someone close will die. However, the elderly find it hard to make new friends as old ones die, so they may turn in on themselves and become lonely. Another cause of loneliness is when families move further away; though loneliness can still be part of life for old people, even when their families live nearby.

It is likely that one of the causes of feeling alone in old age is deafness. It is very difficult to hold a conversation with someone who appears to ignore you, so many of us give up, not realising that the old person has not heard us. It is important to have a great deal of patience to overcome the frustration of talking to someone who is hard of hearing. Patience is often further stretched when the old person takes a long time to understand what is being said – another result of the brain slowing down. Even once we have adjusted to the physical problems of talking to the old person who is hard of hearing, some of the topics of conversation may become irritating especially to the young. Phrases such as, 'When I was young', 'Things were better in the past', and 'The youth of today', may be a standing joke, but are sadly based on fact. It is important to remember that old people's short-term memories may be poor, but past memories are still vivid to them.

Assignments on ageing: feelings and attitudes

1. In pairs, with guidance from your teacher, research one of the following:
   a) Find some information about families in third-world countries. You will probably find many of the facts you need in the sociology section of the library. One of the terms used to describe this type of family set-up is the 'extended family': look up this reference in the library catalogue, and select a few useful books.
   b) The extended family existed in this country until the last century and, according to research carried out by Willmott and Young, continued in the East End of London until the 1950s. Find out about the extended family in Britain in the past. You will find many of the facts you need in the sociology and history sections of the library.

   In the same pairs, present your findings to the group in the form of a short talk.
2. Individual work. You have now read the section 'Another side of old age' and have researched into facts about the extended family. In a short essay (about one and a half sides of A4), discuss the problems of the elderly in Britain today. You may wish to include some of the following points:
   - Have the problems of the elderly today always been the same?
   - What are the particular problems associated with old age in recent years?
   - How do you feel the elderly cope?
   - What can we, as individuals, do to lessen these problems?
   - Is there anything else society should be doing to help?

Ill health

It is obvious that, as the body ages, more things are going to go wrong with it. We are rather like cars in the sense that things generally run well at the start but gradually function less efficiently with age. However, with a little extra care taken during our lives, the health problems associated with age could be reduced in many cases.

Mental-health problems

Depression The onset of depression may be hardly noticeable at first – perhaps a slight loss of appetite and general interest in life, along with disturbed sleep. The individual may believe he or she has a serious physical illness, which may make the symptoms worse. The person feels flat, empty, and useless – and often too low in spirits even to seek medical help. There are many causes and types of depression but, if the symptoms are treated early enough, there is a greater chance of recovery.

Confusion This state was once linked with senility, and thought to be incurable. Nowadays, it is seen in many cases to have a physical cause, such as poor nutrition, glandular malfunctions, or infections, etc. Occasionally, social conditions may cause the elderly to lose touch with reality, and they may imagine they are being persecuted by others. Again, this problem can be treated. In most cases of confusion, the person lives in a world of his or her own and is not in touch with everyday life.

Medical problems

Heart failure This state occurs when the heart is temporarily unable to pump enough blood around the body. The first symptoms are shortness of breath and fatigue, later followed by swollen ankles. If a doctor has not been called at this stage, then the symptoms will get worse. It is important that all problems to do with the heart are dealt with immediately.

High blood pressure The symptoms of high blood pressure vary – from fatigue, breathlessness, and chest pains on the one hand, to no symptoms at all. In old age, higher blood pressure is quite normal, as it ensures that the blood is pushed up to the brain. If, however, any existing symptoms get worse, then a doctor should be contacted. It is a good idea to prepare ourselves before reaching old age: we can lessen the likelihood of high blood pressure by eating a nutritious diet and taking regular exercise (see chapter 21).

Thrombosis If our arteries are not smooth and healthy, but full of deposits, it is possible that clots of blood may form from time to time. If a clot, or thrombosis, develops in the coronary artery (the artery supplying blood to the heart muscle), the person will suffer what is known as a 'heart attack'.

The symptoms are, usually, a pain in the chest which lasts for some time and may spread to the arms. The patient should remain resting until medical help arrives.

A clot of blood that passes through the blood circulation to the brain will cause a stroke. The result of this is that a part of the brain will become temporarily or permanently out of action as it has been starved of the vital blood supply. It is commonly thought that all strokes lead to paralysis and/or impairment of the senses, but this is not necessarily so. Some strokes may be mild and are only detectable by changes in behaviour, such as difficulty in thinking, talking, walking, or writing, or a general lack of interest etc.

A major stroke, however, does cause problems. One side of the body is frequently paralysed, and there may be speech loss. With a good rehabilitation programme, it is hoped that many stroke sufferers will regain a good deal of their independence.

Clots elsewhere in the body are less severe, but need to be carefully monitored in case they move to the heart, brain, or lungs.

One way to lessen the possibility of developing a thrombosis is to encourage the elderly to take plenty of exercise, in order to keep the blood circulating. It is even possible for the bed-bound to exercise with the help of a trained physiotherapist.

Treating high blood pressure reduces the number of strokes.

Arthritis Arthritis is a very painful complaint which, although it can attack any age group, usually develops with ageing. The term literally means 'inflammation of the joints'. There are two main types of arthritis:

1. *Osteo-arthritis* This usually develops in the larger joints, such as the knees and hips, etc. It is caused by wear and tear on the joints, a process usually associated with ageing. It need not necessarily be very painful, but it will cause swelling and restriction of movement in the joints. The symptoms can be reduced in two ways: by the sufferer's losing weight, which will put less strain on the joints, and by regular exercise to strengthen the muscles supporting the joints.
2. *Rheumatoid arthritis* This form of arthritis can develop at any age, but is more common in late middle age. The symptoms may start with inflammation and fever, or fatigue and loss of appetite and weight. The joints will become painful and swollen. Again, the symptoms can be reduced if the sufferer loses weight and exercises regularly. There are various medications on the market that aim to reduce symptoms, but there can be side-effects, and as yet there is no cure. However, medical and surgical treatment, probably through the hospital rheumatology department, can give great benefit.

Baldness Although in most cases baldness is not the sign of any illness, it may have harmful psychological effects on the individual. It affects men more then women and tends to run in families. In some cases, however, hair loss can be a sign of anaemia or thyroid defects.

Hernias A hernia is when part of the body pushes through a weak part of the muscular wall. It can occur in various parts of the body, but the following two are usually associated with old age:

1. Rupture, which is when part of the bowel pushes through the abdominal wall, is often brought on by physical exertion, such as heavy lifting, etc. It is seen as a bulge in the groin and should be reported to the doctor. An operation is usual. If the patient is unfit or has to wait for the operation, the symptoms will be relieved by wearing a surgical truss.
2. Hiatus hernia is where part of the stomach has pushed through the muscular wall of the diaphragm. The symptoms are usually felt as a sensation of burning indigestion in the chest area, due to stomach acid affecting the oesophagus when the patient is lying down or stooping. The patient is usually encouraged to lose weight if necessary and to wear loose, comfortable clothing. In more severe cases, the hernia will need an operation.

Incontinence Bladder incontinence, which is when the individual loses control of the bladder, is relatively common in the elderly and usually suggests that there is something else wrong with the sufferer, such as an infection or, in a man, prostate-gland trouble. If it cannot be remedied, it is possible to reduce the size of the problem by seeking advice about the care of the incontinent from a health visitor: there is plenty of specialised equipment available.

Constipation This is a problem for all ages, but may increase with old age. Constipation, apart from being very uncomfortable in itself, may cause haemorrhoids (piles) and other conditions. Many people believe they are constipated if they do not open their bowels every day, but this is not so. Constipation is when the bowel motions are hard, dry, and difficult to pass. A good diet, with plenty of roughage and liquids, coupled with regular exercise, should help reduce the likelihood of constipation. If the symptoms do develop, there are many good mild laxatives available from the chemist's.

Rheumatism When our muscles or joints stiffen and ache for no apparent reason, we tend to call it 'rheumatism', or sometimes 'fibrositis'. The attacks may be brought on by many factors, including the weather, bad posture, or stress. At the onset of pain, a mild pain-killer such as aspirin should help relieve the symptoms. The problem may increase with age, as the joints and muscles are then more prone to aches and pains.

Assignments on ageing: ill health

1. A certain number of medical terms referred to in this section may need some additional explanation. As a class, check that the following terms are fully understood: high blood pressure, thrombosis, hernias and ruptures, constipation, rheumatism, arthritis, depression and senility, incontinence, hypothermia, haemorrhoids.
2. See if you can arrange for someone to come and talk to the group about old age. There are various people and organisations who could be approached and, if they cannot help directly, they might be able to recommend someone else. Some people to try are:

   - Health Education Council (HEC). It will have films and leaflets available free of charge, and may help find a speaker for you.
   - Social-services department. It frequently has a social worker with special responsibility for the elderly, and is also responsible for local-authority day centres and homes for the elderly.
   - Community education, who may occasionally run special services for the elderly, such as art therapy and handicrafts.
   - Voluntary agencies, such as WRVS, Help the Aged, and Age Concern, often help the elderly in the community and may be able to give advice.

   Decide beforehand what questions you may want to ask the speaker, and check whether he or she requires any special equipment, such as a video recorder, overhead projector, projector, etc.
3. There are films or videos available about old age. Ask your local Health Education Council about borrowing them to show to the group.
4. In pairs, find out about the following conditions and report your findings back to the whole group in the form of a short talk. Check that each pair is researching a different topic, and that each topic is covered at least once.

   - depression
   - confusion and senility
   - arthritis
   - incontinence
   - hypothermia

   With all these topics, it is important to relate your findings to the elderly. Include the following in your research:

   - symptoms and possible causes of the problem
   - remedies or cures available
   - how the problem may affect the elderly – physically, practically, and emotionally

Care of the elderly

When talking about the elderly, it is important to remember that they have the same needs as we all do. They have a right to lead as independent a life as possible, surrounded by their own possessions and friends. This is ideal as long as they are healthy and able to cope, but problems may develop that endanger them, and then steps have to be taken to ensure their safety. However, at present almost 90% of old people do live in their own, or rented, accommodation.

Once it has been accepted that old people can no longer cope on their own, it is necessary to decide what is best for them. It is important that this decision is made, as far as possible, by the old people concerned. Even though they may no longer be able to cope alone, they still have the right to make decisions about their future. There are many choices available, but the decision depends on the needs of each individual.

The home
It is commonly thought that the best place for old people is with their own families, but this is not always the case. Nowadays, houses tend to be smaller, with no spare room for granny or grandad to live in. In the past, it was the woman of the house who had the responsibility of caring for the elderly relative, but now she is as likely to be out at work as the man, and not at home to care for an invalid. Also, with the availability of housing alternatives, many families feel that the responsibility of caring for the elderly relative is too great. This responsibility is even greater if the old person requires extra care.

All this shows just one side: it is quite possible that the old person may be unhappy to move in with the family and does not want to be a burden and tie to them all. The person may feel that living alone, even in a local-authority home, is a preferable alternative.

Sheltered accommodation
Sheltered accommodation, which usually takes the form of warden-controlled flatlets, is provided by both the local authority and voluntary housing associations. In most cases, the flats are in a custom-designed building, where the needs of the residents are taken into account. There are lifts, ramps, intercom systems, and so on. The tenants are independent and have their own furniture and possessions in self-contained flats. Occasionally, they may have to share the bathroom facilities, but this is becoming less frequent. There is also a communal lounge and television room for those who like company. The presence of a warden 24 hours a day means that, in an emergency, the residents always have someone on hand who can be summoned at the press of a button.

This arrangement is absolutely ideal for the old person who is healthy enough to maintain his or her independence, but as yet there are not enough places available to satisfy the increased demand.

Homes for the elderly
These can be run by the local authority, by private organisations, or by voluntary agencies. Local-authority homes are provided to care for those who are too old or infirm to look after themselves, and are run by the social-services department. As with so many local-authority provisions, the standard of the homes varies, even within the same county. Some homes take only those who are still fairly independent, whereas others will take the incontinent and physically or mentally handicapped. In the assignment following this section, you will be asked to find out about the provision of local-authority homes for the elderly in your area.

Privately run homes need to make a profit, unlike those run by the local authority. The standards of privately run homes are checked by the local authority and, if it is not satisfied, it will withdraw its certificate of registration. Due to the large workload of the social-services departments these days, it is not possible for privately run homes to be checked very frequently, but most of these homes do offer a reasonable standard of care, and they tend to take fewer residents than council-run homes. It is important to remember that expense does not necessarily guarantee good care, however.

Homes run by voluntary agencies are, again, smaller than those run by the local authority. They may be run by any of the following:

- religious denominations (this does not necessarily mean that the residents have to belong to the denomination concerned)
- charities dealing with the elderly
- organisations catering for various handicaps, such as blindness and deafness, which organise homes for the elderly handicapped
- professional bodies, which offer homes for retired employees

Geriatric units and hospitals
If the old person concerned has a long-term medical problem, it is likely that the doctor will refer him or her to a geriatric unit, which is specially designed to deal with the medical needs of the elderly.

The long-term aims of geriatric units are either to treat the old people and then return them to their own homes, or to care for those who are no longer able to care for themselves. A good unit will encourage patients to be as independent as possible, while caring for their medical needs.

As yet, these units are not available throughout the country. However, in some parts of the country it is possible for old people to pay regular visits to a day hospital where they can receive treatment during the day, returning home at night.

For families caring for an elderly relative at

home, there is the option of short-term hospital care, where the old person is admitted to hospital on a regular basis. This allows the patient to receive medical treatment, and it also relieves the family of some of the pressures of caring for the sick old person at home.

In areas where these options are not available, ill old people will have to be admitted to wards in general hospitals – usually psychiatric, orthopaedic, or medical wards, depending on their symptoms. This is not an ideal situation, as over one-third of all hospital beds are now occupied by the over-75s.

Assignments on the care of the elderly

1. As a group, carry out the following tasks. In the section you have just read, it is clear that standards and requirements vary from home to home, and county to county. Even within your local authority, the services offered by different institutions will vary.

   You will find the local-authority homes for the elderly in your area listed under the social-services department in your local telephone directory. Make a list of the homes near you, and decide on two or three you would like to visit. Base your choice, if you wish, on some of the following factors:

   - Is the home in the town, city, or country?
   - Is the surrounding area predominantly middle-class or working-class?
   - Is the home large or small?
   - How well-cared-for is it from the outside?
   - Do the residents look happy and occupied?

   Try to get as good a range of contrasts as possible, which will make your research more interesting.

   Once you have made a decision, ring up the home and ask if you can make a visit. Explain why you wish to come, and the staff will be more likely to give you additional help and information. Before you go, devise a list of points you would like to look out for at each of the homes. These points will be used as the basis of a comparative study on homes for the elderly in your area.

   Remember to talk to the residents as well as the staff, but be careful not to appear rude or critical of the home. You are there to collect information, not to inspect or comment on the home, which could arouse the hostility of the staff. Some questions to include are:

   - Is the home adapted or purpose-built?
   - When was it built?
   - How many residents are there?
   - What is the average age of the residents, and how old are the oldest and youngest?
   - How far away are the shops?
   - Are residents encouraged to go to the shops at all?
   - If not, is there someone available to go shopping for them?
   - Is transport available?
   - Is there a lift in the home?
   - Is there a communal room for the residents to use if they wish?
   - Are the bedrooms shared?
   - If so, how many to a room?
   - Do the residents have their own possessions and/or furniture?
   - How far are the bathrooms from the bedrooms?
   - Are there steps etc. to negotiate to get to the bathrooms?
   - Is there any specialised equipment in the bathrooms to help the infirm to bath, shower, or use the toilet independently?
   - What night care is available, e.g. are there alarm-bells by the bedsides, etc.?
   - Are the fire precautions adequate? Would there be any difficulty in reaching the fire exits if necessary?
   - Are there double rooms for married couples?
   - How many staff are there, and what is the ratio of staff to residents?
   - What are the duties of the staff?
   - How many part-time staff are there?
   - What times are the meals, and are the times at all flexible?
   - What is a typical menu, and is there a choice?
   - Recreation facilities
     - what is there for the residents to do during the day?
     - are there regular social activities, such as bingo, films, outings or a trolley-shop?
     - are there other visitors apart from families?
   - Can friends and relatives visit whenever they like?

- Are there facilities for hairdressing, dental treatment, and chiropody, and do the people concerned call on a regular basis? Are these services free of charge?

The following questions may need to be researched from your local authority after the visit:

- Financial arrangements. Find out the charges made to the residents at each of the homes. How do these charges affect their pensions and benefits?
- Admission. How does your social-services department deal with applications to enter homes?

2. Contact your local medical social worker (in the past, they were called 'almoners'). Medical social workers are usually to be found at the local hospital, and it is part of their role to deal with the problems of old people.

   The social worker will be able to inform you about the homes that are available for the elderly in your locality, including the provision of geriatric units. Before you contact him or her, decide what questions you would like to ask.
3. So far, much of your work has dealt with the elderly in institutions, where they have lost some of their independence. Here you are asked to see what is available in the community to help the elderly retain as much of their independence and self-respect as possible, in their own homes. Find out what is available in your area to help the elderly in the following situations. You may need the help of your teacher to direct you to the information.

   a) Is there a home-help service available? If so:

   - How does a person qualify to get a home help?
   - Is a charge made?
   - What duties is the help expected to carry out?

   b) Is there a day centre? If so:

   - Is it free of charge?
   - Is transport available?
   - Is the transport free of charge?
   - What activities are offered there?

   c) Are meals-on-wheels provided in your area? If so:

   - Who provides them?
   - How much do they cost?
   - Are they available to anyone who wants them?
   - What time do they arrive? (It is a common complaint that they are too early for many people.)

   d) Are there any clubs for the elderly? If so:

   - What are they?
   - What time of day do they operate?
   - Do they operate all the year round?
   - Is transport available?

   e) Is there a Good Neighbour scheme operating in your area, or any other type of community support scheme?

   f) Is there a volunteer agency operating in your town centre? If so, what services does it offer the elderly at home?

   g) Is there a holiday scheme operating in your area? If so:

   - How does an old person qualify for a holiday?
   - What is the charge?
   - Who operates the scheme?

   h) If an old person becomes disabled in any way, but still wishes to remain independent, is it possible to hire equipment from the local authority (items such as wheelchairs and commodes, etc.)?

   - What equipment is available?
   - Is it possible to have hand-rails and specialised kitchen equipment installed?

   i) Is there any help available towards home nursing?

   - Can equipment be borrowed?
     - Is it free of charge?
     - Is there a time limit?
   - Is there a laundry service for the incontinent?
     - Is it free of charge?
   - Is there any form of home-nursing provision available, such as a night nurse or a district nurse?

   j) Is there a mobile library service, and are there large-print books for those with poor eyesight?

k) Emergency telephones:
- Can the elderly have a telephone installed in their homes free of charge?
- Is there any financial help towards paying the bills?

l) Travel:
- Are concessionary fares on public transport offered to the elderly in your area?
- If so, how does the elderly person qualify for a pass?
- Does the pass allow the holder to travel at any time, or is it limited to certain hours?

Some places to go for information: the social-services department, Health Education Council, British Red Cross Society, various charities for the elderly (Age Concern, Help the Aged, etc.), various charities for the handicapped (Royal National Institute for the Blind, Royal National Institute for the Deaf, etc.), Women's Royal Voluntary Service, and citizens advice bureaux.

These are only some ideas – there are bound to be some more contact points in your area.

Financial support

So far, we have dealt with the physical and social needs of the elderly in our society, both for the healthy and the less healthy old person. Money, however, can either create problems or help to reduce them.

Many old people complain of lack of money and of how they are unable to cope with the increased cost of living. In most cases this is to be expected as the elderly person no longer earns a wage. In the United Kingdom, the Department of Health and Social Security is responsible for paying state pensions to those of retirement age, and supplementary benefit to those whose incomes drop below a certain level. Additional income comes from private pensions and insurance schemes, which the individual has contributed to during his or her working life.

Here, briefly, are the various sources of income available to the elderly:

State retirement pension Assuming that the elderly person has kept up-to-date with National Insurance contributions, he or she is eligible for a State pension. Men receive the pension at the age of 65, whereas women receive it at 60. This is often felt to be unfair, as women tend to live longer than men, but the idea dates back to the time when married women were usually supported by their husbands. The intention of the Government is that the State pension should reflect, as far as possible, any increase in national earnings.

Widow's financial support If a woman is widowed, her pension may, in part, be based on the contributions of her late husband. If she is over 60, then she will draw her own pension plus half her late husband's pension. If she is under 60, she will be able to draw a widow's allowance for the first 26 weeks. After 26 weeks, if she has dependants, she can draw a widowed mother's allowance, but this situation is obviously quite rare. Any widow over the age of 50 will inherit her late husband's pension.

Pensions for the over-80s Anyone of this age will not have paid the necessary National Insurance contributions, so is entitled to a pension which does not rely on contributions from the individual.

Attendance and invalid-care allowances The attendance allowance is intended to help those requiring partial or continual supervision, due to certain health problems. However, the old person needs to have been in this condition for six months before it is paid.

Those who spend virtually all their time caring for an elderly invalid in the home are eligible for an invalid-care allowance. This is paid only if they are not receiving any other benefits in connection with the invalidity.

Supplementary pensions These pensions are intended to bring an elderly person's income up to a certain weekly level. This level is based on what are considered to be certain basic necessities, such as the day-to-day living expenses of food and rent, special dietary allowances (for those with special needs, such as diabetics) etc. Grants may be available for furniture, clothing, and fares for hospital visits.

Rate and rent rebates As with all members of society, the elderly may be eligible for rent and rate rebates if they are in receipt of supplementary benefit.

Transport concessions, etc. Some local authorities make arrangements for the elderly to travel on public transport either free of charge or at a reduced rate. British Rail operates a scheme on a national level. Some authorities will also offer evening classes at a reduced rate to pensioners. It is also worth investigating local shops, as many offer reductions to the elderly.

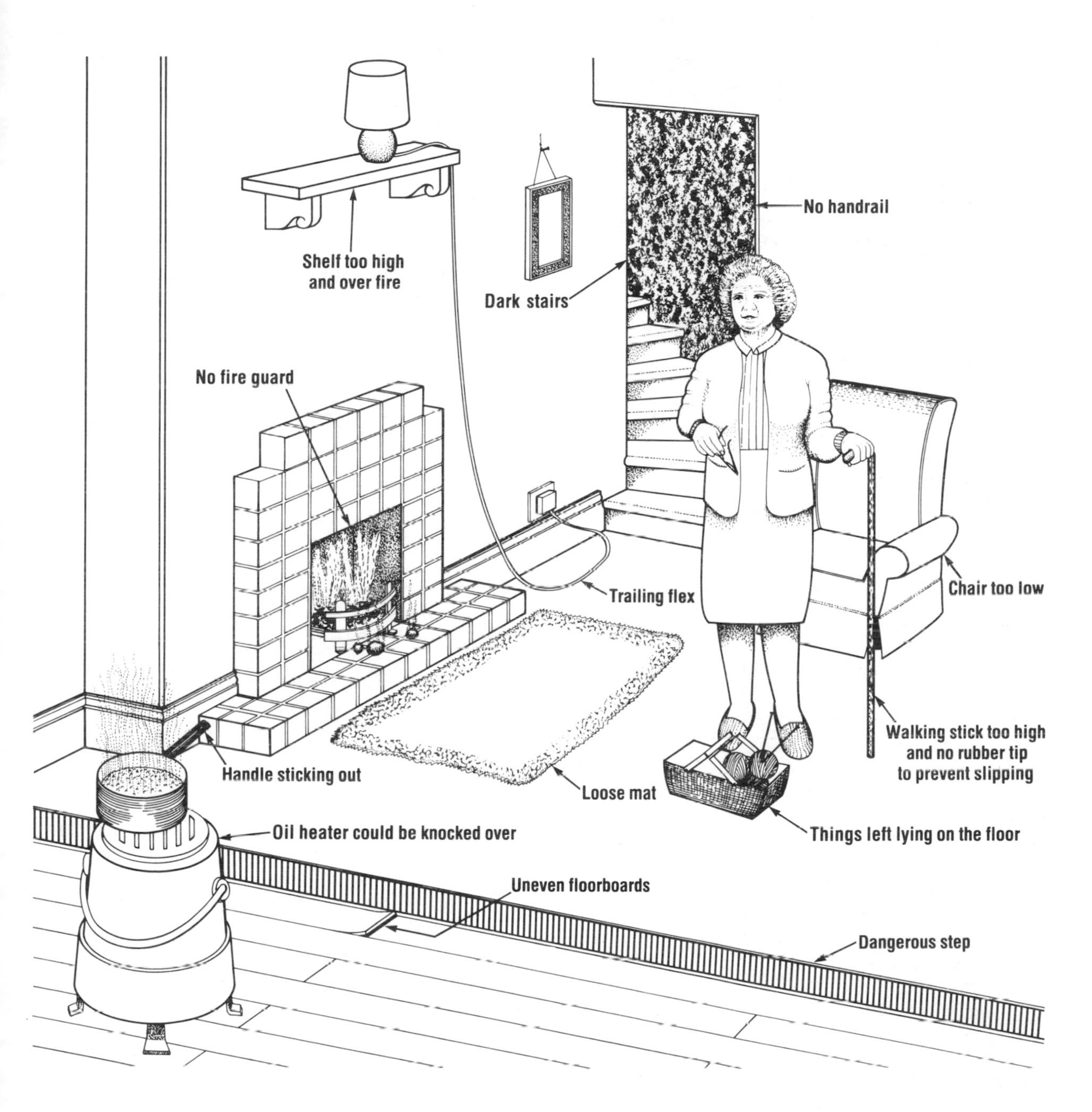

Fig. 15.1 Hazards in the home

Assignments on support for the elderly

1. We have only dealt very superficially with the financial aspects of ageing in this section. As with so many financial situations, the facts vary from person to person, and it would be an impossible task to itemise all the options available in the scope of this book.

   In pairs, investigate the financial help available to the old people in your area. This may come under three categories – national, local, and private help:

   - help offered nationally for housing and heating; help for the sick; free charges for prescriptions, dental, and eye care; etc.
   - help available throughout the county for travel, etc.
   - help offered by local shops – grocers and hairdressers, etc.

2. As a group, it would be useful to invite someone to come and talk to you about the intricacies of the benefit system. Contact your local Department of Health and Social Security who, if it is unable to supply a speaker itself, may be able to recommend someone who knows about the system.
3. Much of the material written about old age deals with the elderly in need. However, as mentioned at the start of this chapter, about 90% of old people live independent lives in their own homes. It is important not to become pessimistic about growing old, especially if you are thinking about working with old people.

   As a group, discuss the whole issue of old age and the ageing process. Relate your views and opinions to the needs of the elderly today, and the needs of those working with the elderly.
4. In pairs, devise a safety plan for old people living in their own homes. List all the potential safety hazards in the average home, and work out how to lessen the dangers. Look at fig. 15.1 to give you some ideas (although there are many more dangers illustrated here than you will find in any one home!).

   Your safety plan may take the form of a series of instructions, a diagram or poster, or a combination of the two. Pin your work on the wall for the group to see.

16 Death and bereavement

Although death is the one experience that all people share, it is also the one experience that we know little about. In the past, death was commonplace: virtually every family suffered the death of a child, and people tended to die younger. In more recent times, death has become an embarrassment: we no longer know how to cope with other people's grief.

The death of an infant

The death of a child is particularly painful, as a child is young and should have a whole lifetime to look forward to.

Recent research has shown that after a stillbirth, the mother can cope with her grief better if she is allowed to hold her dead baby. Many parents find great comfort in meeting other parents who have had stillborn babies, as they have been through the same experience and can give support and understanding.

Cot deaths are still a mystery: no one knows the cause, although there are certain factors that increase the likelihood of cot death. Parents of cot-death babies feel particularly guilty, as they wonder if there was something they could have done to prevent their baby dying. Again, they can find support in other parents who have suffered the same problem.

Many children and adults die unexpectedly as a result of accidents, or illness, war and famine, or suicide. However, as old age draws on, the elderly realise that death is coming nearer. Many of their friends may already have died, so old people are familiar with bereavement.

The dying person

Unless death is sudden and unexpected, the dying person and those around him or her will usually be aware that death is near. People who are dying will usually have lost weight and have very little energy. They will tend to sleep a great deal and be unaware of much that is going on around them. If they are being nursed at home, or in hospital, they need special care to see that they are as comfortable as possible. Relatives may want to be with the dying person towards the end. Although this will be a great strain on them, it will help them come to terms with the death.

After the death

After death, the body is taken away by an undertaker. A doctor must sign a death certificate, saying what he or she thinks was the cause of death. If a doctor has not seen the person before death, the coroner needs to be informed, and he will decide whether to issue a death certificate or recommend a post-mortem examination. If the death occurred in suspicious circumstances, the coroner may decide to carry out an inquest which will decide the cause of death.

Once the death certificate is issued, the relatives can arrange the funeral or cremation.

There is a great deal of work to be done by the relatives of a dead person: in addition to the funeral, the will has to be dealt with, usually through a solicitor. As well as this, there are the deceased's personal effects, and possibly home, to be disposed of. All this comes at a time when the relatives are probably least able to deal with it, as they will still be shocked and upset.

The importance of mourning

We will all suffer a bereavement at some stage in our lives. Research has shown that, if we are able to mourn properly, we will then be better able to come to terms with the sense of loss. Bereavement is one of the biggest crises of human life, and scores high on the stress scale. In general, most bereaved people follow a similar pattern of feelings.

First, they will be shocked and unable to believe the person has really died. Their emotions will be numb, and they may refuse to believe that the person is really dead: this is called the 'denial' phase. Alternatively, they may need to talk about the deceased as if he or she were still alive – another sign that they cannot accept the fact that the person has died.

The burial or cremation is often particularly important to bereaved people, as it symbolises that death really has occurred. Within about a week of the burial or cremation, they feel the grief very badly: many describe it as a physical pain. At this stage, it is essential that bereaved people should not try and 'put on a brave face' – bottled-up feelings are positively bad for them. They will wonder why the death has happened and yearn for physical comfort.

After this stage of grief, bereaved people may begin to feel guilty, perhaps wondering whether they had always done the right things, whether they could have been kinder, and so on. They may idealise the dead person in their minds, and remember him or her as being always better than perfect.

As time passes, bereaved people go through various other phases. Some feel very strongly that the dead person is nearby and, in time, this feeling may be very supportive and secure, as it gives them additional strength to carry on. As even more time passes, bereaved people will realise that they are feeling independent again, and they have come to terms with their loss. They are now ready to carry on with the process of living for the present.

Suicide

Suicide is when a person kills himself or herself because he or she feels that life is intolerable. Such people may kill themselves in a number of ways, including taking a drug overdose, hanging, cutting their wrists, throwing themselves from a great height, and so on. To the rational, healthy person, some of these methods seem difficult to understand: how could somebody throw himself or herself in front of a train? Yet, although many suicides are carried out by people suffering from deep depression and mental anguish, some suicides are planned in a very calm and rational way.

Suicide in a young person is tragic, as young people should have full lives ahead of them. Sometimes their depression and hopelessness about life is so great that there seems little point in living. Their suicide attempts are not always successful, and sometimes they are caught in time to be resuscitated. In many cases, the suicide attempt may have been 'a cry for help' and, with psychiatric help and support from friends and family, the person may find a point in life again.

'Exit' is a society which believes that people who are terminally ill, or in pain, have a right to die if they express a wish to do so. Exit will offer help and advice to such people and their families. This 'voluntary euthanasia', as it is called, is accepted in some countries, where the authorities turn a blind eye to such cases. This type of suicide wish is not necessarily made by a deranged mind: the person may be very rational, and have the full support of his or her family. Aware that their deaths may be long drawn out and painful, such people would rather die in control of their faculties.

Many people worry that if voluntary euthanasia is permitted by law, the next step will be euthanasia of all old people, the handicapped, and so on.

In our society, death is kept very much hidden away. Most of us no longer go to church, and so do not see death as the beginning of life after death. In general, apart from perhaps making a will and taking out a life-insurance policy, we tend to push death out of our minds until we can no longer ignore it. Nowadays, most deaths occur in hospital; whereas, in the past, people tended to die at home with their families. The old are put together in either homes or flatlets and do not often live with their families. All this pushes death away from us. Television, showing deaths on the news followed by deaths in films and plays, leads us to confuse reality with fiction.

Assignments on death and bereavement

1. Find out about funerals:
   - How is a burial or cremation arranged?
   - What is an undertaker?
   - What is 'laying out' the body?
   - What happens during a burial?
   - What happens during a cremation?
2. Write a letter of sympathy to an imaginary aunt whose husband has just died
3. As a group, discuss why you think society today is unwilling to talk openly about death. Why do we use words such as 'passed away', 'passed on', etc. to hide the harsh reality?
4. How do you think children should be told about

a death in the family? Should they just be told that the dead person, or pet, has 'gone away'? Or should they have death explained to them and be allowed to join in with the burial arrangements?

5. What do you think about allowing these people to die if they and/or their family wish it?
   - a terminally-ill cancer sufferer
   - a severely handicapped baby
   - an accident victim, now a 'cabbage' on a life-support machine
   - a senile or bed-ridden very old person
6. Do you think that, by allowing voluntary euthanasia, the next step is to kill all people that society might regard as 'burdens', such as the old and the mentally handicapped?
7. Do you think that donor cards are a good idea? Should we carry cards only if we do not wish our organs to be donated, rather than if we do?

17 Special needs

Physical and mental handicap

Although most babies grow and develop into healthy adults, others are born with a handicap. A handicap may also occur at any stage in life, as a result of an illness or accident.

There are basically two areas of handicap:

1. Physical handicap, which affects the body and prevents it either growing or developing normally, e.g.
   - cerebral palsy
   - cleft palate and hare-lip
   - spina bifida
   - blindness
   - deafness

     (blindness and deafness) These are called 'sensory handicaps', as they affect the senses.
   - deformity
   - handicap as a result of accidental injuries
2. Mental handicap, where a brain abnormality results in a low intelligence. Often, a person with a mental handicap may be physically handicapped too. There can be many causes of mental handicap, e.g.
   - brain damage to the foetus in the uterus, e.g. as a result of rubella
   - brain damage during birth
   - brain damage as a result of a head injury
   - Down's syndrome

Congenital handicap

Physical and mental handicaps which are present at birth are called *congenital handicaps*. These may be immediately obvious, as with a cleft palate, congenital hip dislocation, and most cases of Down's syndrome or they may only show up later, when the child fails to develop at a normal rate.

There are three causes of congenital handicap:

1. It may be inherited from the parents' genes, e.g. Down's syndrome, muscular dystrophy, cystic fibrosis, and haemophilia.
2. There may be malformation of parts of the foetus's body during pregnancy. These may have no apparent cause, as in cleft palate, hare-lip, congenital hip dislocation, and hole in the heart, or they may be the result of the mother's catching rubella, smoking, or drinking alcohol.
3. Neurological problems – damage to either the brain or the spinal cord – will cause congenital handicap. *Cerebral palsy* – commonly known as spasticity – is caused by damage to the

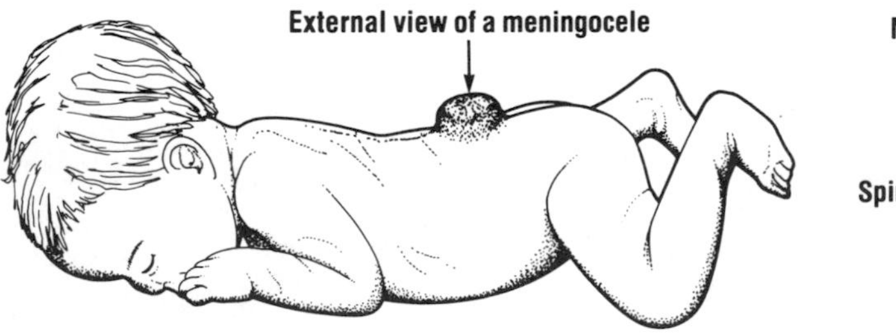

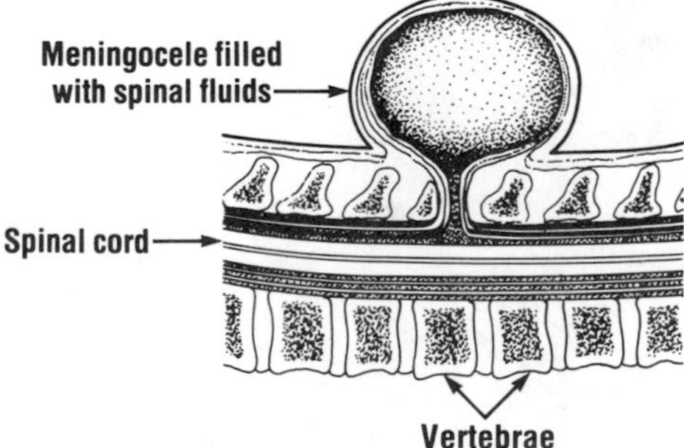

Fig. 17.1 Spina bifida

developing foetal brain. *Spina bifida* (fig. 17.1) is a malformation of the spinal cord, thought to happen early in pregnancy, which causes varying degrees of handicap.

Handicap as a result of disease or accident

Many children and adults who were born perfectly healthy will develop handicaps later, as a result of an accident or a disease. An injury to the head may cause brain damage, injury to the spine may lead to paralysis, and injuries to other parts of the body may cause permanent damage. Diseases such as polio may cause paralysis, and meningitis can lead to deafness or mental handicap.

The physically and mentally handicapped adult

The physically and/or mentally handicapped adult needs special help in order to cope with everyday life. The parents who had helped him or her during childhood and adolescence will be growing older and will be less able to cope with the demands of a handicapped adult. Many handicapped adults will not be at work, so poverty may be a problem.

Although some handicapped people live in homes, warden-controlled flats, or hostels, most live at home. For the physically handicapped, the home will often need adapting, and a wheelchair and/or adapted car will be needed.

There is help available within the community. Day centres for the physically and mentally handicapped offer recreation and training; the health service will provide a home help and home nurse if needed; and some voluntary organisations will offer 'sitting' services, holidays, and so on. However, life is far from easy for handicapped adults and their families.

Learning difficulties

Many children, although not obviously handicapped, have certain learning difficulties at school. Some will find reading particularly difficult, whereas others will have problems with numeracy or spelling.

The child with any of these problems needs special help, based on a thorough assessment of his or her needs. Sadly, many children's needs are not met, either because their problems pass unnoticed by parents and teachers, or because there is inadequate funding for remedial education.

Speech and language development

Speech and language development are known to play an important part in a child's intellectual development. We use language to communicate our ideas to others, to question the world around us, and to think. Without it, educational achievement at school would be very limited. Some children may be backward in speech because they have been deprived of language at home: with extra teaching, they can soon make up for lost time. Other causes of late language development may be emotional problems or intellectual retardation, which can be helped by the right treatment.

Speech therapists may be able to help a child overcome disorders such as stammering or speech impediments. Occasionally, physical malformation of the tongue and soft palate will affect language development. A cleft palate will need to be repaired before speech can be normal, and the extra-long tongue of a Down's syndrome child may make normal speech difficult.

The final cause of speech problems is brain malfunction. A child with receptive dysphasia will be unable to distinguish sounds, and a child with expressive dysphasia will understand speech but find it difficult to organise a meaningful reply.

The disadvantaged child

Handicap is usually regarded in terms of being physical or mental, but there are two other large areas of handicap that affect the child's development: *social disadvantage* and *emotional maladjustment*.

Children who are socially disadvantaged show no obvious signs of handicap, yet the deprivation may be enough to affect their development. Another difference is that social deprivation comes from outside the children rather than within, from the homes they live in and the families around them.

The factors leading to social deprivation are:

- low income
- poor housing
- poor nutrition
- unemployment
- poor parental attitudes

It is obvious that often some, or all, of these factors may appear together. If the father is unemployed, there will be a low income and poor housing, and often poor nutrition. Statistics show that children coming from this kind of environment are more likely to fail at school and become delinquent teenagers. As adults, they will become inadequate parents, and the cycle of deprivation will carry on. Many experimental studies have been carried out to try to find ways of compensating for the deprivation and, sadly, it seems that there is very little that teachers or social workers can do to make a lasting improvement.

Maladjustment, as with the other forms of handicap, will affect a child's ability to learn at school and get on in life. Emotional disturbance and maladjusted behaviour, as well as affecting those around the child, also cause the child to be unhappy and insecure in herself. There is no typical behaviour of children with this problem, but they may show their unhappiness and insecurity by:

- bed-wetting
- becoming withdrawn
- becoming violent
- stealing
- psychiatric disorders

The causes of the problem are similar to those for social deprivation: a poor home environment, plus poor parental attitudes, plus occasionally child abuse.

Mental illness

Mental illness is frequently confused with mental handicap, but mental illness is an illness of the mind. As with physical illness, the more severe the illness, the longer it takes to recover, but many cases of mental illness are cured after treatment.

Mental illness affects a person's thoughts and emotions, perception, motivation, and ability to communicate. Mental illness can have many causes, and is more likely to occur as the person grows older. However, it is generally accepted that mental illness is brought about by some sort of life-crisis situation which the person cannot cope with. The causes can be separated into six groups:

1. Family relationships may put an intolerable strain on some people.

   A child may become disturbed because of the home background – perhaps there has been a lack of affection and love – and this may develop into mental illness as the child grows up.

   Teenagers may find home life and parental attitudes, together with the various stresses of adolescence, too much for them to cope with.

   In adulthood, marital and sexual problems may lead to states of anxiety and distress.

   Doctors may not have time to deal with the problem but will often suggest psychiatric help, or marriage-guidance counselling if the problem lies in a complex relationship.
2. Environment, as in emotional and social handicap, plays an important role in a person's life. Poor housing, poverty, unemployment, or even being in the wrong job, may put intolerable pressures on a person who is unable to adjust to the demands being made on him or her.
3. The mind may be affected by alcohol or drug dependence, brain neuroses or head injuries. Each of these will affect the brain cells and lead to various personality changes.
4. Mental illness may occasionally be triggered off by physical illness, such as flu or glandular fever. Childbirth and the menopause may upset a woman's hormone levels to such an extent that she becomes depressed.
5. Gender appears to affect mental illness, as three women are likely to be admitted to hospital for every two men. An alarmingly high proportion of young women at home with children suffer from clinical depression.
6. Age plays a part in mental illness: the older you are, the greater the chance of developing a mental illness. Old age brings about certain mental problems – particularly dementia, which is associated with a gradual deterioration of intellect and personality. One of the major symptoms of dementia is forgetfulness of recent events, whereas the distant past is easily remembered. 'Confusion' is when dementia, coupled with other problems, such as poor vision and hearing, leads to the old person's losing touch with reality.

Mental illness takes many forms, but it can be roughly divided into two types: *neurosis* and *psychosis*. Of these two, neuroses are the longer lasting, less serious conditions, which are easier to treat. Usually the person is aware of the problem and understands the way in which he or she is reacting. Psychoses are more serious, as the individual is out of touch with reality and withdraws into his or her own world.

Anxiety neuroses, in which the person has deep-seated fear and anxiety about various conditions, are the most common form of mental illness. Those affected fear for their health and even develop symptoms that have no real cause, or they may develop anxieties about work or school.

Phobias, another form of neurosis, are fears that have got out of proportion – such as fear of spiders, open spaces, and so on. The person knows that these fears are irrational, but cannot overcome them.

Psychoses are far more serious, and psychotics cannot be reasoned with as they do not realise that their actions are abnormal. Psychotic disorders, such as manic depression and schizophrenia, are relatively uncommon.

Professional help for the mentally ill

Within the health service, there are many people qualified and trained to help the mentally ill. The initial referral will probably come from a GP or social worker, and then the psychiatric health-care team will take over treatment. Treatment will take place on an outpatient basis for less severe mental illness: more severe cases will be admitted to a psychiatric hospital for treatment.

Treatment for the mentally ill

Treatment may take the form of drugs, psychotherapy, aversion therapy, or electric shock treatment. After treatment, the person needs to become part of the community again. This rehabilitation process needs careful handling in order to reduce the possibility of mental illness recurring.

Assignments on special needs

1. Intelligence is measured by finding the Intelligence Quotient (IQ). Find out about IQ:
   - What is the average intelligence?
   - What is the IQ of a mentally handicapped person?
   - What is the IQ of a genius?
   - How is IQ measured?
2. Check you understand these terms: congenital, genes, genetic.
3. It is very distressing for parents to give birth to a handicapped child. However, it may help them to cope with their problem if they can understand their child's handicap and find out what help is available.

   Below is a list of mental and physical handicaps. Each member of the group should research into at least one of these handicaps. Include these points:
   - whether the handicap is congenital or acquired through disease etc.
   - possible causes
   - possible symptoms
   - possible treatment

   Handicaps
   - autism
   - anencephaly
   - blindness or partial sight
   - cerebral palsy
   - cleft palate
   - club foot
   - coeliac disease
   - congenital hip dislocation
   - cretinism
   - cystic fibrosis
   - deafness or partial hearing
   - Down's syndrome
   - haemophilia
   - hole in the heart
   - hydrocephalus
   - epilepsy
   - multiple sclerosis
   - muscular dystrophy
   - phenylketonuria
   - poliomyelitis
   - Siamese twins
   - sickle-cell anaemia
   - spina bifida
   - spinal cord or brain injury
4. Learning difficulties are often difficult to measure or assess. Find out anything you can about these two disorders:
   - developmental aphasia
   - dyslexia

   Find out also what your area provides for children with special educational needs.
5. Special schools are for handicapped children. They offer more individual care to the children and help prepare them for their future lives.

   Find out about your nearest special school. Try to arrange either a visit to the school or for a member of staff to come and talk to the group.

   How does it differ from ordinary schools?
6. Social deprivation is considered to be a handicap, as it affects the child's ability to develop into a fulfilled adult. It may be caused by these

factors:

- low income
- poor housing
- poor nutrition
- unemployment
- inadequate parents

Write a brief article explaining how these factors can cause a child to be socially handicapped.

7. Emotional disturbances and maladjusted behaviour are caused by similar factors. Here are some behavioural problems: bed-wetting, stealing, truancy, vandalism. As a group, discuss how these problems may develop in a child.
8. The child guidance centre offers help, assessment, and counselling for the parents and the child. Find out the work done by your nearest CGC. Perhaps a member of staff, or a social worker, would be able to come and talk to the group about the work he or she does.
9. There are many statutory agencies whose job is to help the handicapped. Check that you understand, in brief, what each of the following agencies does for the handicapped.

   a) The health service provides:

   - general practitioners (GPs)
   - health visitors (HVs)
   - school medical officers
   - speech therapists
   - occupational therapists
   - physiotherapists
   - psychologists/psychiatrists

   b) The social-services department provides:

   - generic social workers
   - social workers attached to CGC

   c) The education services provide:

   - teachers in special education
   - education welfare officers
   - educational psychologists

10. There are too many voluntary agencies to list here. Obtain a copy of *The Directory of Associations* from the reference library and look up the names of some voluntary agencies dealing with handicap. Choose one that interests you, and write off to the organisation, enclosing a stamped addressed envelope for further details.

    Present these details as a brief fact sheet that can be photocopied for the rest of the group, to create a dossier of voluntary organisations dealing with the handicapped.
11. There are many aids available to help the physically handicapped cope with their disability and lead as normal a life as possible. Find out about these. Information is available from:

    - The British Red Cross (address and telephone number in the local directory)
    - The Disabled Living Foundation, 380–384 Harrow Road, London W9 2HU

    If writing, please send a stamped addressed envelope. The DLF also allows visits, so, if you live near London, it might be worth arranging a trip.
12. Obtain a copy of 'Help for handicapped people' (leaflet HB1) from the DHSS, and answer these questions:

    - What is the difference between a contributory and non-contributory benefit?
    - Look up attendance allowance. Do you think it is fair that married women had not hitherto been eligible for this benefit? What problems would this cause?
    - Using this leaflet, find out what benefits and other help is available for:
      - a mother of three young children, disabled with multiple sclerosis, who has financial support from husband, but would like to be mobile
      - a partially sighted child
      - a physically handicapped father, now out of work through handicap, who wants to know everything he is eligible for
      - an elderly grandad, living at home with unmarried daughter, who is unable to walk due to war injury

13. In most areas, there is an adult training centre or sheltered workshop, where handicapped school-leavers and adults can go every day. The trainees, as they are called, do simple contract work for local industry, which makes them a useful part of the community. In many cases, the centres will also have other training input, such as literacy and numeracy, cookery, etc.

    Try to arrange a visit to your local centre to

see how it operates.

14. Most areas will have a day centre, where physically handicapped school-leavers and adults can attend on a regular basis. Try to arrange a visit to your local centre, and find out the following:

- Who comes to the centre?
- How often do they come?
- Is transport available?
- What is offered at the centre (e.g. hairdressing, chiropody, bathing, speech therapy, etc.)?

15. Find out about these mental illnesses:

- depression
- schizophrenia
- paranoia
- hypochondria
- psychosomatic disorders
- phobias: claustrophobia
  xenophobia
  agoraphobia, etc.
- manic depression

Write briefly on each, explaining symptoms, causes, and treatment where possible.

16. The Mental Health (Amendment) Act 1983 lays out ways of dealing with various types of mental condition. Obtain a copy of the Act (published by HMSO, 1983), and look at Sections 2, 3, 4, 5, 7(guardianship), 136, 37, 41, 35 (removal to hospital), 36 (removal for treatment), 38 (internal hospital order).
17. The psychiatric workers in the health service are:

- psychiatric nurse
- psychiatrist
- psychiatric social worker
- occupational therapist
- psychologist
- psychoanalyst
- disablement resettlement officer

Briefly, find out the role of each of these people. If possible, invite one of them to come and talk to the group about the work he or she does.

18. Find out what facilities are provided in your area to help the mentally ill recover. There may be:

- a mental hospital
- a health centre offering psychiatric counselling
- a psychiatric unit in a general hospital
- psychiatric outpatients clinic – for treatment and advice
- a day hospital – patients return home at night

When you have found out what is available in your area, try to arrange a visit.

19. Mental illness can be treated in various ways. Find out briefly what each of the following entails:

- drug treatment
- psychotherapy
- aversion therapy
- ECT

20. Senile dementia and confusion are states of mental illness associated with old age. Find out the symptoms and causes of each of these conditions.
21. Mental illness has an effect on the relatives of the person concerned. In small groups, discuss the effect the mental illness of these people may have on their immediate families:

- a teenage daughter who is suffering from anxiety neurosis about school
- a young mother who has developed severe post-natal depression after the birth of her baby
- a middle-aged father of two teenagers who has developed schizophrenia and thinks he can hear voices
- a middle-aged woman who refuses to leave her house as a result of agoraphobia
- an elderly grandmother, living with a daughter and her family, who has become confused and is convinced the family are plotting to kill her

18 Drug abuse

Drug abuse is the term applied to the non-medical use of drugs – that is, the use of drugs not prescribed by a doctor to cure an illness or condition. Although drug abuse is in the news a great deal today, man has taken mind-altering drugs as far back as history records, and alcohol and tobacco are socially acceptable drugs in our society. However, even drugs used as medicines are open to abuse.

With all drugs, repeated use may lead to addiction. In extreme cases, addicts will do virtually anything in order to obtain the drug – even if it means altering their moral code.

Drug dependence is the term used to describe the state a person has reached when the body shows signs of withdrawal after the drug is stopped: A smoker will suffer withdrawal symptoms, although they may not be so obvious as those suffered by the heroin addict.

We need to note why a person becomes dependent on certain non-prescribed drugs. Firstly, pressure from friends seems quite high on the list of reasons. The need to conform to the behaviour of friends and acquaintances – the term used for this is *peer-group pressure* – is high for teenagers, and if your friends are taking drugs, it is more likely that you will experiment with them. With many adults, drugs are used to calm the nerves and as a means of escape from the problems of everyday life. Some people may feel depressed or bored, and see drugs as a way of brightening up their lives.

Non-prescribed and prescribed drugs

There is a wide variety of drugs involved and these can be divided into four groups:

1. The first group are the *narcotics* – the so-called 'hard drugs' such as heroin, morphine, and opium, all of which are derived from the opium poppy. These drugs induce sleep and drowsiness, and often a feeling of peace and tranquillity. Morphine is the drug in its pure form, and from this is derived heroin, which is twice as strong as morphine.

   These opium-derived drugs, called *opiates*, have medical uses as pain-killers, cough-suppressants, and for the treatment of diarrhoea. However, there were side-effects, so in an attempt to avoid these, synthetic opiates were made for treating pain: pethidine (widely used in childbirth), diconal, distalgesic, and physeptone. The general term applied to the opiates and synthetic opiates is *opioids*.

   Unfortunately, since the 1910s the illicit use of heroin has increased. However, heroin bought illegally is highly likely to have been contaminated by the addition of powders which look similar.

2. The second group are the *stimulants* – the amphetamines and appetite-suppressant drugs, commonly known as speed, sulphate, and cocaine. These speed up the body's activity to such a great extent that the person never stops moving, and needs no sleep.

   Cocaine is found naturally in the South American coca shrub. The South American Indians have chewed the leaves as a stimulant since time immemorial, but it was not until the mid-nineteenth century that cocaine became available as a medical drug and was used to treat a wide variety of ailments. At the same time came cocaine abuse. The medical profession did not find cocaine the great cure-all they had originally hoped for. Until the 1920s, cocaine was an ingredient of Coca-Cola and various tonic wines, as it acted as a stimulant. Recently, there has been renewed interest in cocaine amongst certain elements of the population.

   Amphetamines are synthetic drugs originally developed to treat depression and for appetite-suppression. Nowadays, they are prescribed for people suffering from narcolepsy (a condition where the individual falls asleep during the day) and, strangely, for hyperactive children.

3. The third group of drugs are the *hallucinating drugs*, such as LSD, cannabis, and psylocibin, which produce strange sensations and a false sense of understanding.

   Cannabis is derived from the Indian hemp plant and its widespread use is as a relaxant, rather like alcohol and tobacco. 'Hashish', or 'hash', is obtained from the resin at the top of the plant. 'Grass', or 'marijuana', is the crushed leaves, flowers, and twigs of the plant. The strength of each of these varies, but generally hash is the stronger.

   Many people feel that as taking cannabis is apparently no more dangerous than smoking tobacco or drinking alcohol it should be legalised, whereas others argue that the long-term effects are not yet known, and why add another potentially dangerous drug to the two already legally available? Research has shown that, as most marijuana is smoked, it carries the same health risks as smoking. However, there are also suggestions that regular use of cannabis may lead to a decrease in fertility, a suppression of the body's natural immunity defence, and brain shrinkage.

   Lysergic acid diethylamide (LSD) is a synthetic hallucinogenic drug which is taken to induce 'trips', in which the individual's senses are heightened and time seems to slow down. The effect of the drug is very much dependent on the individual's personality and the surrounding environment, so someone who is depressed, or in a state of anxiety, is more likely to have what is called a 'bad trip'.

   Mescalin, which is found in a South American cactus, is another hallucinogenic drug, but it is not widely available as a non-medical drug. Another hallucinogenic drug is *psylocibin*, found in the Liberty Cap mushroom.

4. The fourth group are the hypnotic drugs which put you to sleep.

   Barbiturates were until the 1970s prescribed as sedatives in order to calm people suffering from anxiety, and as sleeping pills to help those with sleeping problems. In small doses, these drugs kept the person relaxed, but in large doses they could prove fatal. Tuinal was the most commonly abused of these drugs. In many ways, barbiturates are similar to alcohol, and indeed many users also had alcohol-related problems. The withdrawal symptoms are similar to those of alcohol withdrawal – anxiety, shakiness, and sometimes convulsions and delirium.

   Tranquillisers are prescribed to help control anxiety and stress, and to help sleep problems. The best-known of these are Librium, Dalmane, and Valium. Until recently, doctors have prescribed tranquillisers in ever larger numbers. However, the growing number of adults dependent on them in order to get through the day was giving cause for concern, and doctors now no longer prescribe them so freely. Long-term use can lead to unpleasant side-effects, such as tension, nervousness, sleep disturbance, sweating, sickness, and diarrhoea. Although doctors generally thought these drugs were non-addictive, patients suffered clear and obvious withdrawal symptoms. It is now thought that between 30 and 40% of those regularly taking tranquillisers become dependent.

   Tranquillisers can be valuable treatment for some forms of mental illness, such as schizophrenia, but for many people, they merely temporarily cover over the underlying problem. Being shown how to cope with the real cause of tension, and having someone to talk to, can help some people to avoid the use of these drugs.

Solvent abuse

Solvent abuse by school-age children has become a major problem. Youngsters use various solvents, such as certain glues, cleaning fluids, and fuels, or they may inhale the propellent gases used in aerosols. The feelings of 'drunkenness' and merriment give teenagers a quick, cheap 'high'. In the short term, solvent abuse can lead to vomiting, drowsiness, and even unconsciousness. More serious, however, is the fact that death can be caused by accidents resulting from drowsiness, and inhaling vomit. The long-term effects of solvent abuse can be brain and liver damage.

Alcohol and tobacco

Alcohol and tobacco have long been acceptable social drugs in our society. The USA tried to prohibit alcohol in the 1920s, but found that people then brewed it illicitly.

Governments in Europe have made varying attempts to reduce smoking, although cynics say that governments could not afford to lose the revenue they gather from the taxes. In the United Kingdom, about £80 million is spent by the tobacco companies on promoting their products, whereas only £1 million is spent by the government on anti-smoking campaigns. Doctors say that smoking is the greatest cause of preventable death and disease in the UK.

Smoking

Cigarettes and cigars are made from tobacco, which is the dried leaves of the tobacco plant. Cigarette smoke contains *nicotine*, which is a stimulant, and tar, which is harmful to our bodies.

Some children start smoking at the age of five, and research has shown that people who start smoking in childhood are less likely to give up the habit in later life. By the age of 14, about one-third of smokers were offered cigarettes by their parents, even though most parents would rather their children did not smoke.

Smoking seems a strange habit – after all, what is attractive about smoke? When asked why they smoked, youngsters offered the reasons that smoking had a tough and mature image, and they thought it would attract the opposite sex.

Even though virtually all schoolchildren are aware of the health dangers associated with smoking, this does not seem to bother them. In many cases, the addiction makes it an impossible habit to give up. Research has been unable to come up with an answer as to why some people can give up smoking, even after many years, whereas others become addicted more easily.

There are many potential health hazards in smoking. Apart from the obvious risk of heart and lung diseases, there is an increased chance of a mother giving birth to an abnormal or stillborn baby if she smokes during pregnancy. Apart from these major hazards, smoking affects the senses of smell and taste, and a smoker's cough can later develop into bronchitis.

Given all the facts showing that smoking is bad for health, there must be some advantages, otherwise no one would continue to smoke. Research shows that smokers like the 'kick' they get from a cigarette, they find that getting and lighting a cigarette gives them something to do with their hands, and smoking the cigarette is a form of pleasant relaxation.

So far, we have discussed cigarette smokers, but others smoke cigars and pipes. In general, although the same health hazards apply, they tend to be slightly reduced, as the smoke is not inhaled so deeply. However, a cigar or pipe smoker is more likely to develop cancer of the mouth and throat.

Alcohol

Alcohol in our society is very much a part of our everyday lives. Alcohol is used to mark a celebration, and at social events, as well as to help us calm down and relax at the end of the day.

Alcoholic drinks are made by fermenting grains, fruits, or vegetables. They vary in strength, depending on the alcohol content – beer tends to be relatively weak compared with spirits. Unlike cigarettes, the drug in alcohol is a depressant rather than a stimulant.

Drinking alcohol has caused controversy because of the dangers associated with drinking and driving. A person can be unfit to drive a car safely even after a relatively small amount of alcohol, although he or she may be far from drunk or an alcoholic. Ideally, drivers should drink no alcohol. If they wish to go out for a drink, then they should make alternative arrangements for getting home.

Although there are legal limits for drinking and driving, there are no hard and fast rules about how much alcohol makes a person reach the limit. If you are a woman or a light man, the limit will be reached sooner than if you are a heavy man. Your tolerance level for alcohol can be affected by taking drugs (both prescribed and illicit drugs) and by many other factors – so the safest way is to avoid alcohol completely if you are driving or operating machinery. The difficulty is that, after a few drinks, people tend to become over-confident and are more likely to believe that they are perfectly fit to drive. Sadly, the number of driving offences committed by people over the limit is increasing, and has reached the level it was before the introduction of the breathalyser.

Alcoholism

We tend to have a stereotyped picture of an alcoholic as someone who is completely drunk and walks around holding a bottle. However, only a minority of alcoholics are like this: it would be difficult to diagnose that the majority of alcoholics had a drink problem at all. Alcoholics may drink to make themselves feel better, rather than for social reasons. They frequently drink alone, and they may need a drink in the morning. They may have committed a drink-related offence and, although they will deny having a drink problem themselves, friends and relatives think they do. Alcoholism is present if the drinking gradually increases over a period of time.

Extreme addiction to alcohol, rather like tobacco, may happen relatively quickly or take a number of years, depending on the individual. The addiction is difficult to cure, and the withdrawal symptoms can be painful: fears, sleeplessness, panic, shaking (known as 'DTs' – *delirium tremens*), and even fits. It is a difficult task to give up alcohol at this stage and, even after a long period of abstinence, a drink can bring back the habit within a few days. There are various societies that aim to help the severe alcoholic give up the habit.

For the heavy drinker, alcohol may bring about a pleasant feeling for a short time but, as it is a depressant, the drinker will feel depressed and irritable after a few days. The heavy drinker's relationships will suffer, as he or she becomes either loud-mouthed or aggressive. A great many wife-batterers do so under the influence of alcohol. The alcoholic's sex life will also suffer, as alcohol causes impotence in men. Even social drinkers will find their sex life temporarily affected by an evening's heavy drinking. Physically, prolonged heavy drinking does quite a bit of damage. The most common alcohol-related disease is *cirrhosis of the liver*. The death rate from this disease is used as a measure of a country's alcoholism.

Alcoholics will lose weight and be more likely to suffer from gastric upsets, stomach ulcers, and internal haemorrhage. Their muscles weaken, including those of the heart, which leads to weakness and breathlessness. Even the mental capacities are affected – there is a high rate of brain damage, memory loss, and early senility.

Alcoholism affects not only the individual, but also the people around him or her. To be drunk in public is an offence, and extreme cases may result in prison sentences. More frightening is the fact that a high percentage of violent crimes are linked with heavy drinking. The majority of football hooligans and about half of convicted rapists had been drinking when they committed their offences. Alcoholism can also be measured in terms of money: this country pays out millions of pounds in sick pay for days off as a result of heavy drinking.

The alcoholic's family suffers as a result of the habit. Wives with alcoholic husbands tend to be more disturbed, and the children are more likely to become alcoholics themselves.

Almost all studies on alcoholism deal mainly with men – however, recent research has discovered that there is a difference between male and female alcoholics. In general, the female alcoholic turns to drink later in life, often as a result of depression, serious illness, or a traumatic experience in childhood or later life. She is more likely to be influenced by alcoholism in the family than a man, and will find it harder to 'kick the habit' once treatment has started. Women's bodies are less tolerant of alcohol than men's, so they are more likely to suffer the side-effects of drinking (sickness, dizziness, headaches, etc.). It is thought that this factor is likely to keep the number of female alcoholics fairly stable. (The present sex ratio is 3:1 of men to women alcoholics.)

Over the past 10 years there has been an increasing amount of research carried out on the effects of drinking during pregnancy. People have long known that drinking during pregnancy is inadvisable, and the baby of a heavy-drinking mother is more likely to be born with various defects, but it is only recently that the full effects are becoming known. The term 'foetal alcohol syndrome' (FAS) is used to describe the baby born to a heavy-drinking mother. The features of FAS may include

- small head, short nose and thin upper lip, small eyes and flat cheeks
- low birth weight
- retarded growth
- mental retardation
- birthmarks
- heart murmurs

There are obvious problems in assessing the incidence of FAS. The first is that some of the defects, such as a simple birthmark, are extremely small. The second is how to measure 'heavy drinking'. Some studies say two drinks a day, and others allow eight. What is clear, however, is that drinking in the early stages of pregnancy while the foetus is developing is inadvisable, as the risks are too high to gamble with. But it should also be remembered that there are plenty of alcoholic women with perfectly normal children.

Drug abuse is a difficult problem. Informing youngsters of the dangers to their health, or the risk of prosecution and imprisonment, is obviously not a big enough deterrent. Perhaps there will always be a problem while people are willing to disregard long-term health risks for short-term 'pleasures'.

Assignment

1. Check that you understand the meaning of these terms: withdrawal effects, dependence, addiction, tolerance, peer group, trafficking, junkie.
2. As a group, discuss why you think these people may take drugs:
   - 15-year-old teenager – glue sniffing
   - 26-year-old mother at home with children – tranquillisers
   - 18-year-old – on heroin
   - 24-year-old – smoking cannabis
3. In pairs, find out about these drugs by completing this chart:

Drug	How administered	Likely addicts	Short- and long-term effects	How to give it up	What the law says
Opioids Cocaine Amphetamines Cannabis LSD Solvents					

4. Tranquillisers are acceptable drugs and hitherto have been relatively easy to obtain on prescription. Look at these facts:
   - About 10% of all National Health Service prescriptions are for tranquillisers.
   - Of these, many are repeat prescriptions for which the patient does not even see the doctor.
   - 30 to 40% of users become dependent.
   - A quarter of a million people have been using tranquillisers for more than five years.
   - Valium is the most profitable drug ever made.
   - The majority of users are women.

   Discuss the use of tranquillisers. Do you think they help people in stressful situations? Should doctors be doing more than just prescribing pills?

   Nowadays there is an increasing awareness of the possible dangers associated with taking tranquillisers for more than two weeks. Find out what help is available in your area for people who feel that they are dependent on tranquillisers.
5. These are the physical effects of smoking:
   - bronchitis and emphysema
   - heart disease and coronary thrombosis
   - lung cancer
   - pregnant women who smoke have an increased chance of miscarriage, bleeding during pregnancy, low-birth-weight baby, and placental abnormalities
   - children of smoking parents have a greater chance of developing respiratory infections, bronchitis, and pneumonia

   Check that you know the meaning of each of these terms and, as a group, discuss any action that you feel should be taken to stop people smoking.
6. 'Passive smoking' occurs when cigarette smoke is inhaled by non-smokers in the company of smokers. Find out about passive smoking and its dangers.
7. Studies have found that smokers are more likely to drink alcohol and use illicit drugs than non-smokers. Why do you think this is?
8. Nicotine is the drug in cigarettes. Research the effects of nicotine on the body's nervous system.
9. As a group, devise a questionnaire about smoking. Make two to three copies per member of the group. Ask friends or relatives who are smokers to complete the questionnaire, and analyse your findings. Your questions need to find out:

- when they started smoking
- why they started smoking
- how many cigarettes they smoke
- how long they have been smoking
- whether they would like to give it up
- whether they have ever tried to give up
- whether their health has suffered, and how
- whether they fear the health hazards of smoking
- whether they agree with smoking being banned in public places

Add any other questions you think are relevant

10. Get hold of copies of the leaflet 'That's the limit'. Read through the leaflet and, if you drink, fill in the charts yourself. If not, get a relative or friend to complete it.

    As a group, discuss your drinking habits.

11. Write a set of guidelines to help keep drinking at a safe level. Information can be found in books, from the Health Education Council, and from these addresses (send a stamped addressed envelope):

    - Accept Clinic, 200 Seagrave Road, London SW6 1RQ
    - Alcoholics Anonymous, 11 Redcliffe Gardens, London SW10 9BQ
    - Alcohol Concern, 305 Gray's Inn Road, London WC1X 8QF

12. Find out how drug addicts are helped in your area. The following may be available:

    - counselling sessions
    - community homes
    - substitute drug treatment
    - therapy
    - psychiatric help

    Collect the information together on a fact sheet which would be suitable for distribution in your area.

13. Do you feel enough is being done in schools and colleges to warn people of the risks of taking drugs, and can the media do more? Discuss your opinions and ideas. Perhaps you could arrange for more information to be made available at your school or college.

19 Sexuality and sexually transmitted diseases

Advertising

The media encourage us to be sexually aware. Advertisers use attractive male and female models to encourage us to buy products; airlines employ attractive stewardness to serve food and drinks and welcome us on to the plane; the majority of successful film stars or pop artists are good-looking. The list is endless. Women, particularly, are on the receiving end of this approach. They are under continual pressure from advertisements to look good, use make-up, and be slim in order to attract men. Alternatively, they are depicted as successful mothers running a perfect home, with a husband and two perfect children. It is no wonder that so many women feel failures.

Pornography

Pornography is a rather more extreme form of presenting sex than advertising. In general, it portrays men and women as objects for mere sexual gratification. Some people say that it leads people astray and contributes to rape and sexual assault and it should therefore be banned. Pornography can be 'soft', such as that found in the newspapers, or 'hard', which is usually illegal. It is a matter of personal opinion whether we see pornography as acceptable or not.

Sexual taboos

In terms of sexual relationships, society has created many taboos. These are mainly centred on relationships that are unable to fulfil society's expectations of marriage and reproduction. Society sees a threat in homosexual relationships and extra-marital affairs, although attitudes to both are nowadays more liberal.

Incest and paedophilia

Society has made laws prohibiting sex within the family (incest) and with children (paeodophilia).

Incest creates the danger of inbreeding – when weaknesses tend to be duplicated rather than counteracted by strengths from another family – so there is a strong practical as well as moral reason for this taboo.

Paedophilia is considered wrong and harmful because the child is not in a position to consent to the situation and is an unequal partner in the relationship. Society is generally horrified by instances of paedophilia, and considers that children should be protected from early sexual experiences which may affect them badly.

The general attitude in the UK is that most sexual practices, as long as they are not illegal, are tolerated between consenting adults.

Rape

One sexual situation where one person's feelings are not taken into account is rape – when a man forces a woman to have intercourse against her will. It is a common misconception that rape is something that a stranger does to a woman in a dark alley. However, most rapes are carried out by a man known to the woman.

Some people – even some judges – seem to believe that any woman out alone is 'asking for it'. They think that part of the blame lies with the woman, and this attitude is reflected in the way in which rape is dealt with by the law, when the female victim may be accused of 'contributory negligence'.

Sadly, only as few as 10% of rapes are reported, as many women dread the ordeal that a rape enquiry can cause. The victim has to be physically examined by a doctor to prove that rape did indeed take place, and then again later to check that she has not caught VD or become pregnant. Many women who have reported a rape have said that they feel as if they are on trial to prove their innocence, rather than to prove the rapist's guilt.

Emotionally, the rape victim will be deeply upset. She will feel anger at being violated, as well as guilty and ashamed that it should happen to her. She will need support and understanding from friends and family. If this is not possible, then she may be able to turn for help to one of the voluntary groups which exist in many of our larger cities.

Positive attitudes can help, for both rape victims and women who have not been raped but are aware of the dangers. Women can take courses in self-defence specially aimed at dealing with male aggressors.

Wife-battering

Closely linked with sexual violence against women is wife-battering. This can include mental torture, kicking, attempted strangulation, punching, use of weapons (including boiling water) etc.

It is only in the last 15 years or so, since the Chiswick Refuge opened in 1971, that wife-battering has been recognised. As with rape, many people believe that the wife 'asks for it' and must enjoy it because she puts up with it. This does not explain why the battering happens, and neither does it explain what the wife is feeling.

No one really knows the cause, but wife-battering is thought to result from the stresses placed on men who are unable to cope with pressure. They may be unemployed, poor, in bad housing, etc., and feel powerless against the situation – so they use their superior strength against their wives. Wives, in the same powerless situation as their husbands, may turn their violent feelings towards their children in child abuse, or in on themselves and become depressed.

There are many reasons why these women do not leave home. In most cases, there are children to be supported, and the children would be taken into care if the women left home with them. If she left home without them, her husband might be unable to look after them properly, either because he is at work or because he is simply incapable of giving them a proper upbringing. Most women are financially dependent on their husbands, and the chances of finding a job when they have children and no home are very limited.

Many wives take their families to a refuge and then return home again to the same situation. The sad thing is the effect battering has on the children – research has shown that the sons tend to grow up to be wife-beaters and the daughters to be battered wives themselves. Perhaps t so many battered wives return to homes – it is all they have ever known, ev children.

Sexual harassment

Far less unpleasant and traumatic than rape, but nevertheless a problem for some women, is sexual harassment at work. These women have found that, because they are females in a traditionally male-dominated workplace, they are open to sexual harassment by their male colleagues. This can range from bottom-pinching to, in some instances, rape. Again, many people say women 'ask for it' by dressing attractively etc. but to most women it is upsetting to be seen as a sex object and not as someone doing a job.

Sexually transmitted diseases (STDs)

Over the past 20 years, there has been an increase in sexual activity, probably as a direct result of more effective contraceptive methods. Because a person is likely to have more sexual partners, there has been an increase in the number of sexually transmitted diseases (STDs), which are diseases spread almost exclusively through sexual contact. Sooner or later, a person regularly having intercourse with different partners is bound to come across someone who is infected. Obviously, for people who stay with one partner, the chances of contracting a sexually transmitted disease are greatly reduced.

Gonorrhoea

Gonorrhoea is one of the major STDs, and infects many thousands of people each year. Symptoms show within two to eight days of infection as a discharge from the penis or vagina and a stinging sensation on passing water. For many women, however, there may be no symptoms. If the disease is left untreated, it may spread up through the sex organs and lead to sterility for both sexes. If still left untreated, the disease will spread to the joints and eventually to the vital organs, such as the heart.

Syphilis

Syphilis is a far nastier disease than gonorrhoea, but fortunately rarer. Within two to six weeks of infection, the person develops what is called a 'chancre' – a small hard sore – in the place where the infection started (usually the genitals).

For women this may be the vagina, and for male homosexuals this may be the anus, in which case the chancre may pass unnoticed.

If this stage is left untreated, the sore will disappear, but the disease will be spreading over the body, into what is called the 'second stage'. This second stage usually shows itself as a rash and swollen glands, which, again, could pass unrecognised. This phase will pass and the person may seem well for anything up to 10 or 20 years, and then the third, potentially fatal, stage will appear, which affects the heart, nervous system, and brain, leading to insanity.

Because the symptoms of gonorrhoea and syphilis can pass unnoticed in some cases, it is essential that an infected person tells his or her sexual partners that they may have been infected. The special clinic which deals with STDs does ensure that all contacts are traced and tested for the infection.

Non-specific urethritis (NSU)

NSU is becoming increasingly common nowadays. As the name suggests, it is an inflammation of the urethra with no specific cause. At the beginning, the infection may be confused with gonorrhoea, and a microscopic examination of the discharge needs to be made in order for the correct treatment to be given.

Viral infections

There are three STDs which are transmitted by viral infections: genital warts, genital herpes, and AIDS.

Genital warts These may appear on the genitals one to nine months after infection. In men they are usually visible, but women may develop them inside the vagina, as well on the vulva. Although most genital warts are passed on through a sexual encounter, it is possible that they may develop independently. Recent research links untreated male genital warts with cervical cancer.

Genital herpes This viral infection is reaching epidemic proportions in the USA. Unfortunately, virus infections cannot be cured, although there is some hope that a cure may be found in the not-too-distant future.

The herpes virus causes painful blisters, swelling, and sometimes fever. It comes in two forms: HSV 1, which is the type usually found as cold sores around the mouth, and HSV 2, which is the type found around the genitals. The virus will lie dormant in a sensory nerve and flare up when something triggers off an attack, which will erupt as painful blisters, usually in the same place each time. Attacks seem to be triggered off when the person is run down, either through illness or stress.

When suffering an attack of genital herpes, the person is particularly infectious and should avoid sexual contact. When the virus is dormant, partners will not catch herpes.

Sadly, herpes can be fatal to a baby born to a mother suffering an attack of genital herpes. If the mother does develop an attack when the baby is due, doctors will perform a Caesarian section, which will save the baby.

AIDS

Perhaps the STD most in the news recently, and causing the greatest fears, is AIDS (Acquired Immune Deficiency Syndrome). It is still relatively rare in the UK, although there is a higher incidence of the disease in the USA.

AIDS is caused by a virus, and can be passed on either through sexual intercourse or through contact with infected blood. This last method means that haemophiliacs needing to be injected with the clotting element of Factor 8 are at risk, as are drug addicts using other people's syringes and people receiving blood transfusions. However, nowadays donated blood is tested for the AIDS virus.

About 15 out of 20 sufferers are homosexual men, and four out of 20 are drug addicts. The small remainder of sufferers include haemophiliacs and patients receiving infected blood transfusions. Although findings that AIDS can be passed on to partners through the semen is causing concern, as yet very few healthy heterosexuals are at risk.

Trichomoniasis

Trichomoniasis is a disease, usually sexually transmitted, which mostly affects women. It is caused not by a viral or bacterial infection but by a parasite which is able to swim around in the vaginal secretions. This parasite can live for half an hour outside the body – so this is the only STD that actually can be caught from a toilet seat or an infected towel or flannel. The woman will find that she develops an unpleasant discharge and discomfort. In men, it is thought to account for about one in 20 cases of NSU.

Thrush

Another infection causing a discharge and discomfort is thrush. Again, this need not necessarily be sexually transmitted: it can also be caused by the woman's taking antibiotics, being pregnant, or diabetic, all of which upset the chemical balance in the vagina.

Infestations

The final types of STD are those caused by infestations.

Lice Pubic lice are lice that live in the pubic hair, and any other body hair except the scalp hair. The louse lives on the blood it sucks from the host, and the female lays eggs, called nits, which are firmly cemented to each hair. Pubic lice can be caught by body contact, or from infected clothes or bedding.

Scabies Scabies is caused by a minute mite which burrows into the skin to lay eggs. The burrows look like small bumps, and they are very itchy, especially at night when the person is warm in bed. The mite usually burrows around the genitals and buttocks, wrists and ankles, armpits and breast, and between the toes and fingers. Although it is classed as an STD, other adults and children may catch the infection from close body contact with an infected person.

Assignments on sexuality and sexually transmitted diseases

Note The Health Education Council publishes several leaflets on sexually transmitted diseases.

1. Go through magazines and cut out any advertisements which you feel show how the media portray women either as the objects of men's fantasies or as perfect wives and mothers. Do they ever depict women in any other way?

   Do the same for men. How are men depicted in advertisements? Make posters in small groups to show your findings.

   Finally, as a group, discuss whether you feel that present-day advertisements reflect real life.
2. Paedophilia has been in the news over the past few years, as there is a body of adults who believe that sexual intercourse with children should be made legal. As a group, discuss your views on paedophilia.
3. There are many terms used today to describe people with alternative sexual behaviour. Check that you understand the meaning of these terms: bi-sexual, transsexual, transvestite, sadist and masochist, exhibitionist, peeping Tom. Write a brief paragraph for each.
4. Find out about homosexual and lesbian relationships. How much do they differ from heterosexual relationships? What problems might the couples encounter? As a group, discuss your attitudes to these people.

   Discuss whether you think the way in which our society sees these groups of people is just or unjust.
5. Rape is a crime against women. Many people see it as a result of a society which

   - encourages aggression in males
   - believes that all women have a secret desire to be raped and dominated
   - believes that dominant men are attractive to all women, except lesbians
   - thinks that all women say 'no' when they mean 'yes'
   - believes that women exist to be the property of men

   In small groups, discuss these five points. Can you think of examples that illustrate each of them?

   In your area, find out

   - the number to phone if you have been raped
   - if there are any women's self-defence classes
6. Sexual harassment at work has recently been quite topical. Have you ever been sexually harassed? Write your views on the subject.
7. Recent research has found that there are some cases of 'battered husbands', where a dominant wife bullies a meek husband. This situation is the basis of many comedies. Discuss your views of male/female dominance in a relationship.
8. Find out about AIDS:

   - What are the symptoms?
   - How does the blood used in transfusions and Factor 8 needed for haemophiliacs become infected?
   - Why are male homosexuals more at risk?
   - How may a foetus become infected?
9. Nowadays, it appears that STDs are on the increase and some, such as AIDS and genital herpes, are relatively new to us. Some people say that this is the result of a too liberal attitude to sex in the Western world. As a group, discuss the

following:

- What method of birth control is best for reducing the risk of STDs?
- What are the dangers of 'one-night stands'?
- Does the increase in STDs suggest that we should reconsider our attitudes towards: sex before marriage, changes in sexual partners, homosexuality?
- Do you think there will be a moral backlash to our more liberal era, as has previously happened in history?

10. Make a table to show the symptoms, treatment, and possible dangers of the main STDs:

Disease	Common name	Symptoms		Treatments	Other dangers/ side-effects
		Male	Female		
Gonorrhoea Syphilis NSU Genital warts Genital herpes Trichomoniasis Crab lice Scabies					

20 Contraception and abortion

It is only in relatively recent times that sexual intercourse has become separated from reproduction. In the past, most sexual partnerships resulted in pregnancy, because there was little or no contraception. Nowadays, with more effective contraception, the couple have a choice as to whether to have a baby or not, and most love-making will not lead to pregnancy, even for couples who have decided to have a family eventually.

Contraception has to be the ultimate responsibility of the woman, as it is she whose life would be most affected by an unwanted pregnancy. Yet this does not mean that the woman should make the decision alone: in a happy, loving relationship, contraception should be a shared responsibility.

Contraception is a means of preventing a baby being conceived. This may be done either by preventing an egg being fertilised by a sperm – as with barrier methods and the pill – or by preventing a fertilised egg from implanting itself in the wall of the uterus, as with the intrauterine device (IUD). As yet, there is no fool-proof way of avoiding pregnancy – there is still the possibility of human error, product failure, or a combination of both. However, the success rate increases if the couple are well motivated and take advice about which method is best suited to them.

Contraception has always been a controversial area. Roman Catholics are prohibited by their church from using any artificial method of birth control in their marriage, as sexual intercourse is seen only as a means of creating children.

There are many myths and superstitions surrounding contraception (covered in the assignments), and these have grown up because of a lack of understanding about our bodies and the sexual act.

To get contraceptive advice, the woman – accompanied by her partner if she wishes – should go to either her GP or a family-planning clinic.

Once at the clinic, the woman may be given an internal examination (fig. 20.1) by a doctor. Many

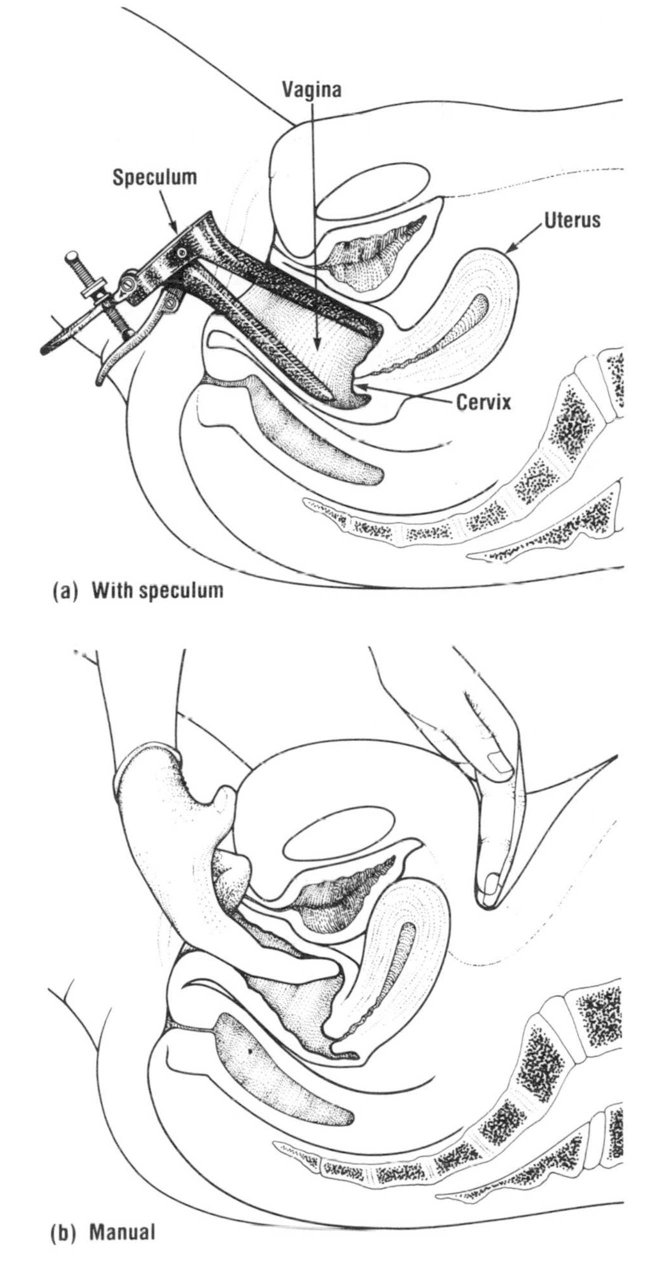

Fig. 20.1 Pelvic examination

women feel worried about this, but there really is no need. A woman who is relaxed will find the examination much easier. Perhaps the knowledge that the examination is carried out to check that she is normal inside and to prevent any problems later will help her to see the positive benefits. During the examination, the doctor will probably take a smear from the cervix to check for possible cancerous cells.

At the visit, other questions will be asked about the woman's background, and she will be weighed and have her blood pressure measured.

These checks will help the doctor to advise her about which method of birth control would be best for her.

Natural methods of contraception

The clinic will recommend the more effective methods of contraception, but some couples rely on less safe and satisfactory methods – the so-called 'natural' methods.

Withdrawal method

In *coitus interruptus* or *withdrawal*, the man pulls his penis out of the vagina immediately before ejaculation. This is a very unsatisfactory method of birth control, but one that is quite widely used. There is the risk that the man will mistime it and, anyway, many sperm are released in the lubricating fluid – before ejaculation – and it needs only one sperm to fertilise the egg. It is also sexually unsatisfying for both the man and the woman, and can lead to feelings of frustration.

Rhythm method

The second natural method is the *rhythm method*, which is accepted by the Roman Catholic church. This involves the woman's finding out her times of ovulation each month by taking her temperature each morning, and then avoiding intercourse for about six days before ovulation and four days after. This method is not particularly effective as ovulation time varies from month to month, and some women seem to have no 'safe period' and can conceive during their monthly period.

Barrier methods

The next group of methods are called 'barrier' methods, because they create an artificial barrier between the sperm and the egg.

The sheath

The most familiar barrier device is the male *sheath* – also known as the *condom*. This is a sheath, made of thin latex rubber, that the man puts on his erect penis. At the top of the sheath is a teat that traps the sperm as they are ejaculated. The man must be careful that the sheath does not come off as he withdraws, and there is always a slight danger that the sheath may break, so it should always be used with a spermicidal cream or jelly to reduce these risks.

Perhaps the greatest advantage of the sheath is that it limits the spread of sexually transmitted diseases – it is said that Casanova used a sheath made of a sheep's bladder for this reason.

The cap

The woman's barrier method is the *cap* or *diaphragm* (fig. 20.2), which closes off the cervix so that sperm cannot reach the uterus. There are various different caps available, and they should be fitted by a doctor, who will decide which size will be most comfortable.

Because the cap cannot form a completely sperm-tight seal, it should be used along with a spermicidal cream.

The pill

The *pill* (fig. 20.3), probably the most effective method of birth control, works by preventing the woman's body from producing an egg. The pill can only be obtained on prescription from a doctor, who will give the woman an examination and check that she is fit enough to take it. She will then need to return every six months for further check-ups to ensure that she has not developed any adverse symptoms.

The advantages of the pill are that the woman is always safe from pregnancy, assuming she has followed the directions, and love-making does not have to be interrupted for some device to be fitted. Many women also like the feeling of being in control of their bodies and knowing when their periods are due. Their periods tend to be lighter and less painful.

The disadvantages are the dangers to the woman's health – especially if she is overweight, smokes, or has a family history of thrombosis.

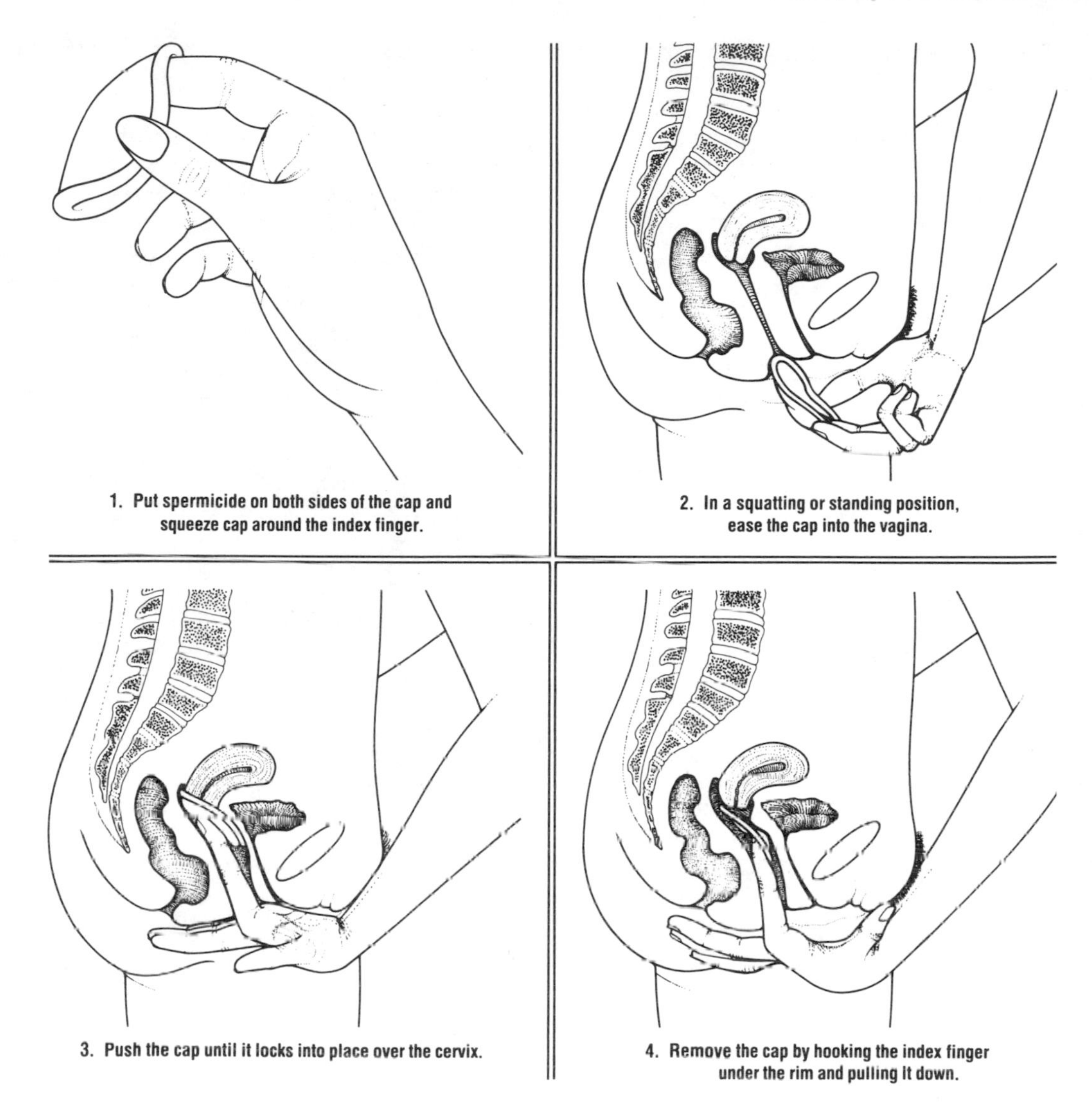

Fig. 20.2 Fitting a diaphragm

The intrauterine device (IUD)

The *intrauterine device* (fig. 20.4) prevents pregnancy by causing the uterus to reject a fertilised egg from implanting itself in the uterine wall. There are many different IUDs, but they all work in the same way. Each IUD is made of plastic, and some have copper built in. It is inserted into the uterus through the cervix and left there for either three, five, or more years. A cord hangs through outside the cervix, so that the woman can check the IUD is still in place. Like the pill, the IUD does not interrupt love-making. It has to be fitted by a doctor to ensure it is inserted correctly and safely, to avoid the risk of pregnancy.

The drawbacks are that some women suffer cramp-like pains during their periods, which may also be much heavier than usual. This usually settles down within a few months, but if it doesn't the IUD can easily be removed. Although the IUD is recommended for women who have had a baby, as the cervix will be wider, there are some smaller versions for women who have never been pregnant.

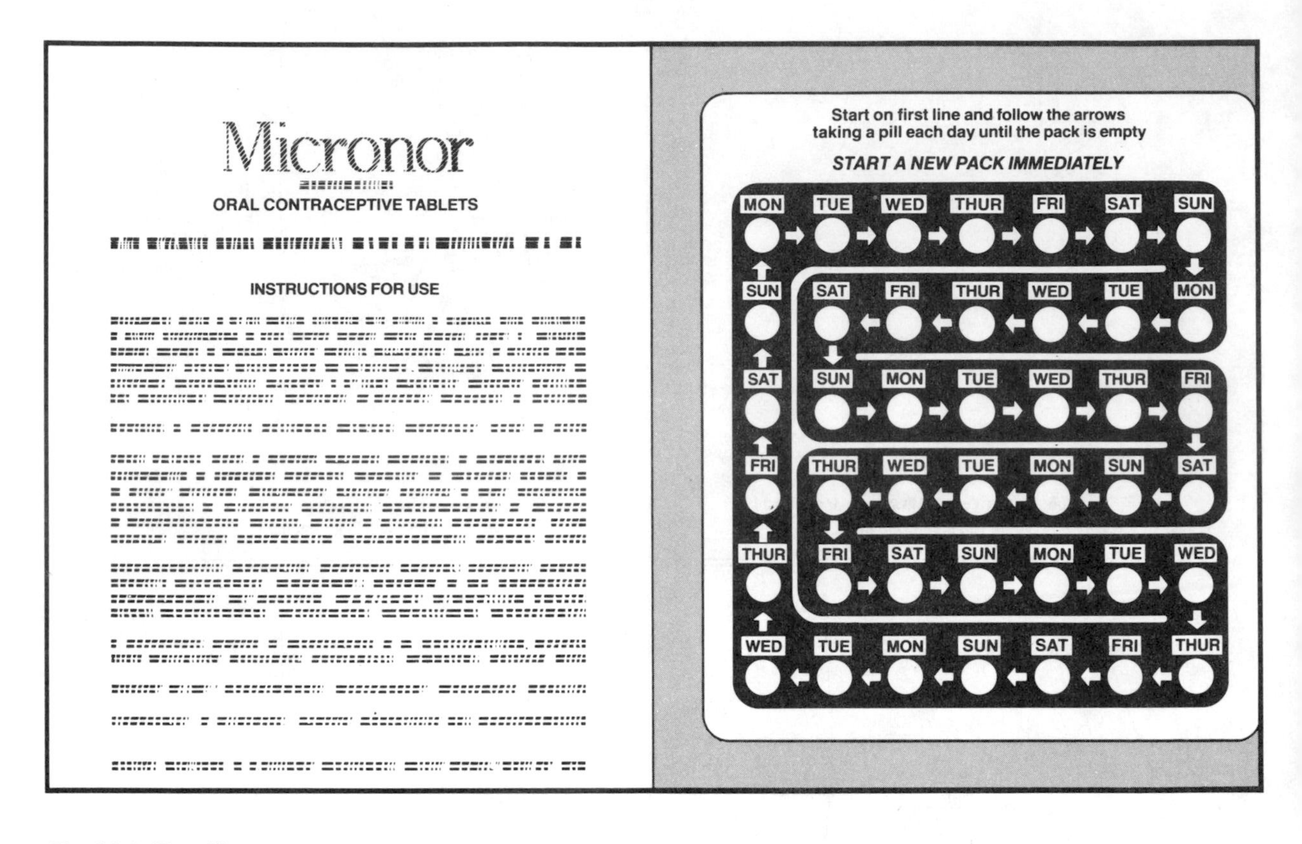

Fig. 20.3 The pill

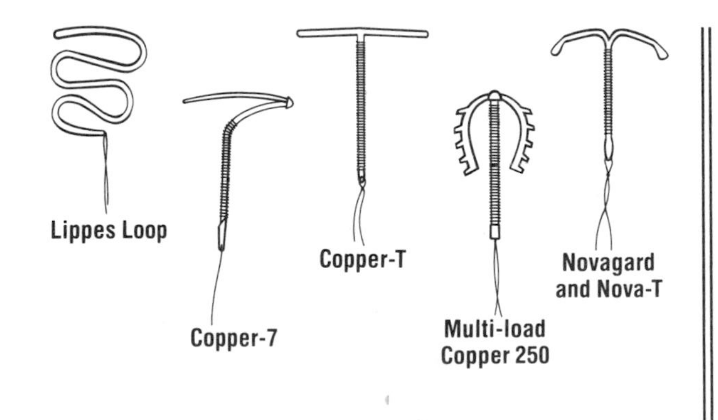

(a) Types of IUD available

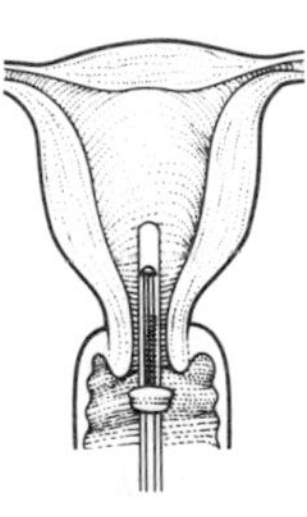

1. The IUD is inserted inside a narrow tube.

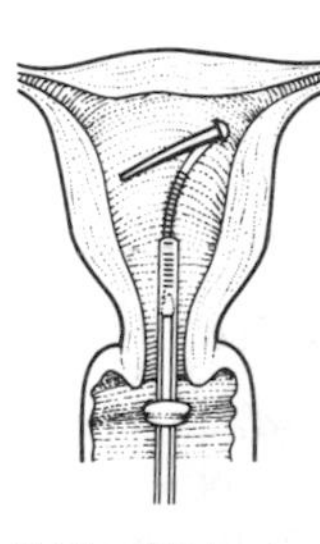

2. The IUD is pushed out of the tube into the womb.

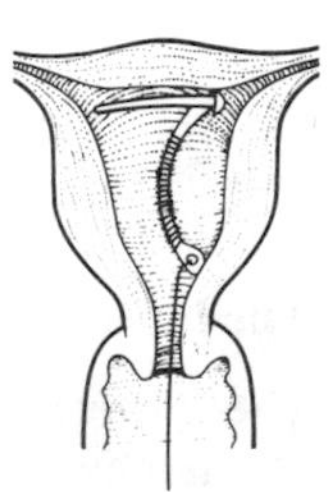

3. The tube is removed and the IUD stays in place in the womb.

(b) Inserting an IUD into the womb

Fig. 20.4 The IUD

New methods of contraception

At present, there are two new methods of contraception available for women.

1. *The Depo-Provera injection* can be given to the woman every three months. Although it is as effective as the pill, many women suffer from nausea. It is, however, a useful method for women who find it hard to remember to take the pill every day.
2. *The contraceptive sponge* is now available. The sponge is impregnated with spermicide, moulded into a similar shape to the diaphragm, and used in the same way. As the man ejaculates, the sponge absorbs the sperm. It must be left in place for six hours and then taken out and washed. Each sponge can be used two or three times.

Other methods are still being tested. These include a hormone pellet, which is implanted under the skin at the back of the neck and releases hormones into the body, a nasal spray, and a hormone-releasing IUD.

Laboratories have been trying to perfect a male pill, aimed at preventing sperm production, but as yet this have proved unsuccessful.

Sterilisation

A permanent method of birth control is by surgical sterilisation.

Vasectomy

For the man, the sterilisation operation is called a *vasectomy* (fig. 20.5). The sperm which are made in the testicles travel along a tube called the vas deferens. In a vasectomy, the vas deferens is cut and the ends are tied, thus preventing the sperm from passing along the tube. As the seminal fluid can still travel along the tube from the gland that produces it, there is no noticeable difference in the man's ejaculation. Neither will intercourse feel any different to him. The only difference is that there is no sperm in the semen, so there is no chance of pregnancy occurring. The sperm are still produced, but are reabsorbed by the body.

For complete safety, the man is advised to use other methods of birth control for about two months after a vasectomy, until any remaining sperm have been cleared out of his system.

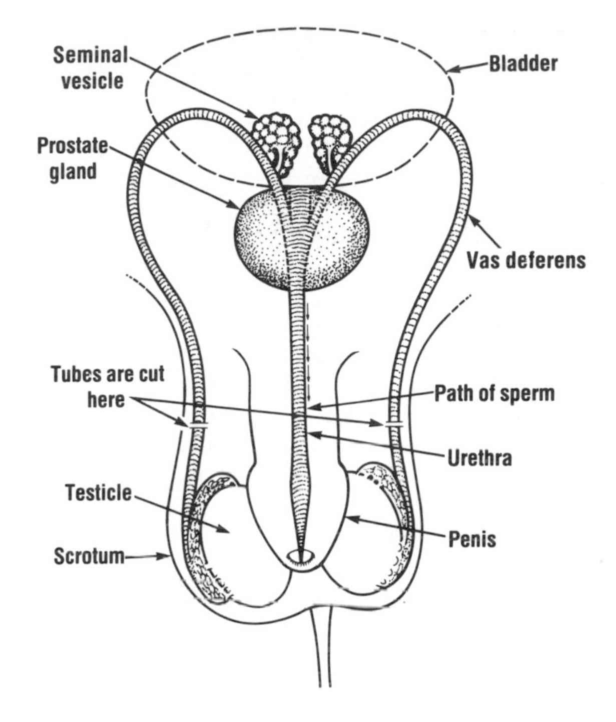

Fig. 20.5 Vasectomy

Female sterilisation – tubal ligation

For women, sterilisation (fig. 20.6) should also be regarded as a permanent method of birth control.

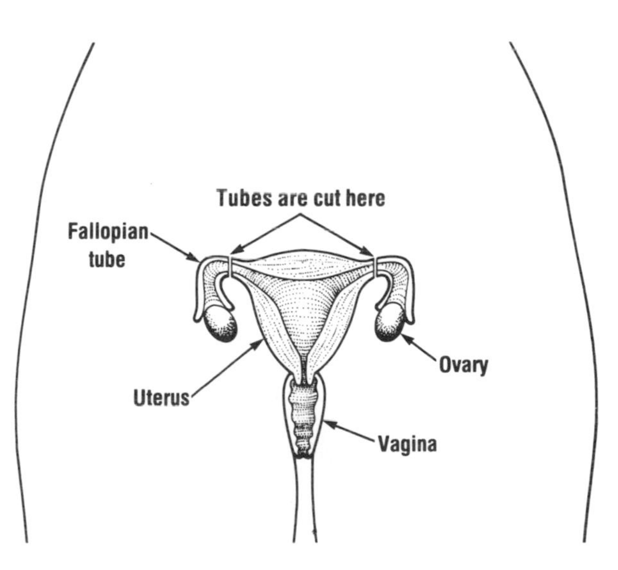

Fig. 20.6 Tubal ligation

Many women misunderstand the operation and fear for their femininity. As you can see from the diagram, the ovaries are left intact and continue to produce all the hormones as usual. What happens is that the Fallopian tubes are cut and tied, in order to prevent the egg from meeting the sperm. The eggs, rather like sperms after a vasectomy, are reabsorbed by the body.

Both male and female sterilisations are simple operations and usually involve less than a day in the hospital or clinic. The man has a local anaesthetic and is ready to go home within hours. The woman usually has a general anaesthetic and so needs a few hours to recover before leaving. It is, however, very important that the couple make a joint decision about sterilisation, regardless of who is going to have the operation. Although it has been reversed in rare cases, it is a permanent method, and couples should see it as such.

Dealing with an unwanted pregnancy

Sadly, for various reasons, not all babies are wanted babies. Occasionally there is a need for methods of preventing a birth after unprotected sexual intercourse.

The 'morning after' method

The 'morning after' method is used within 72 hours after intercourse. There are two methods. Firstly, the woman may be given a high dose of the hormones in the contraceptive pill, which may have side-effects, such as sickness or diarrhoea. Secondly, an IUD can be fitted in order to prevent any fertilised egg from becoming embedded in the uterine walls.

Abortion

Abortion as a means of ending an unwanted pregnancy has been in evidence for thousands of years. Until fairly recently, abortion had always been a 'back-street job', and many women died and even more were made very ill, or sterile, because of the infections they caught. When it was realised that making abortion illegal did not prevent it from being carried out, many countries legalised it. As a result, far fewer women died from abortions. In 1967, abortion became legal in Great Britain, and the number of illnesses and deaths due to abortions has dropped dramatically.

Doctors prefer to do an early termination – that is, before 12 weeks or occasionally before 15 weeks of pregnancy. They use two methods:

1. *Dilatation and evacuation* (D & E), used when the foetus is 7 to 12 weeks old, involves sucking out the contents of the uterus (fig. 20.7).
2. *Dilatation and curettage* (D & C), used when the foetus is 8 to 12 weeks, or occasionally 15 weeks in some cases, involves scraping out the contents of the uterus (fig. 20.8).

It is much safer to have an abortion at this early stage, as a late abortion does carry greater risks. At 16 to 28 weeks, the foetus is too big to be removed by suction or curettage, so labour has to be chemically induced.

Obviously, abortion has a psychological effect on a woman, especially a late abortion. The woman may feel guilty and upset, and experience a sense of loss for the baby. It is important that counselling should be available, both before and after the operation, to ensure that she gets all the emotional support she needs.

Assignments on contraception and abortion

Note The Health Education Council publishes many useful leaflets on contraception. The Family Planning Association, now listed in the local directory, also produces leaflets and is often willing to send a representative to give a talk to a group.

1. Discuss the following myths about contraception, adding any others you may know. A woman won't get pregnant if she
   - doesn't have an orgasm during sexual intercourse
   - holds her breath during sexual intercourse
   - is breast-feeding
   - lies on top of the man during sexual intercourse/has intercourse standing up
   - is under 16
   - passes urine immediately after sex
   - is having a period
2. Find out about Victoria Gillick and her campaign to make it illegal to give under-16s contraceptive advice without parental consent. When you have the information, discuss your feelings about what she has tried to do, from the points of view both of a parent and of a young girl.

   What would be the dangers of such a law?

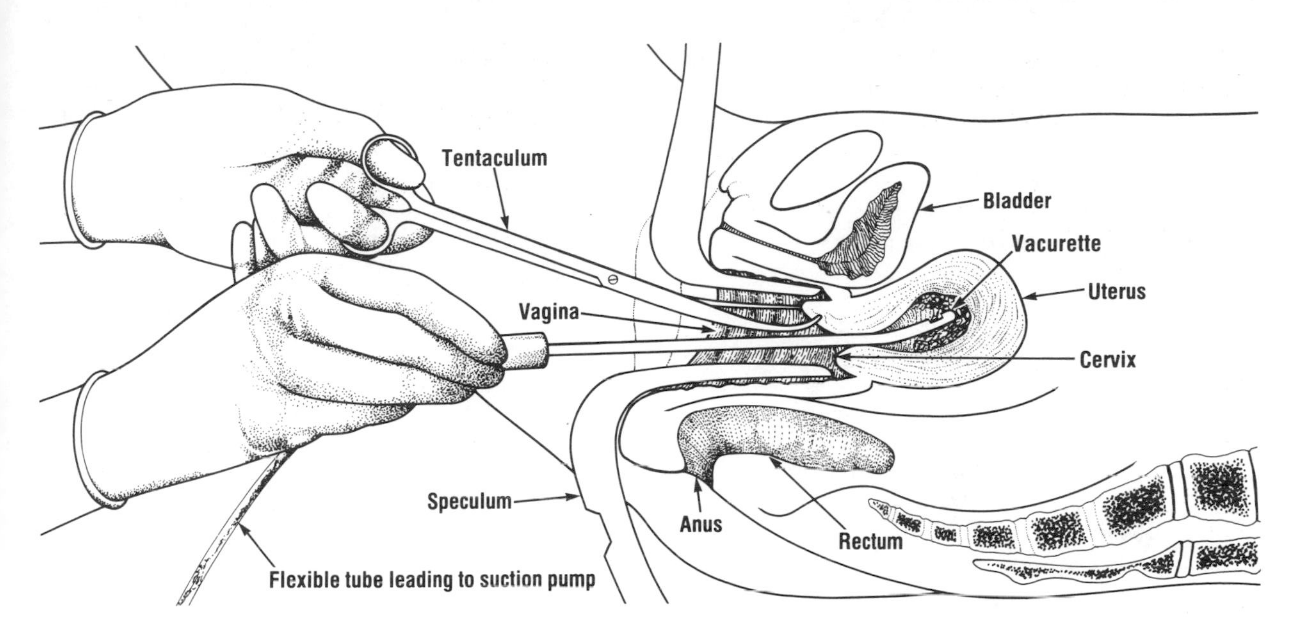

Fig. 20.7 An abortion, using dilatation and evacuation

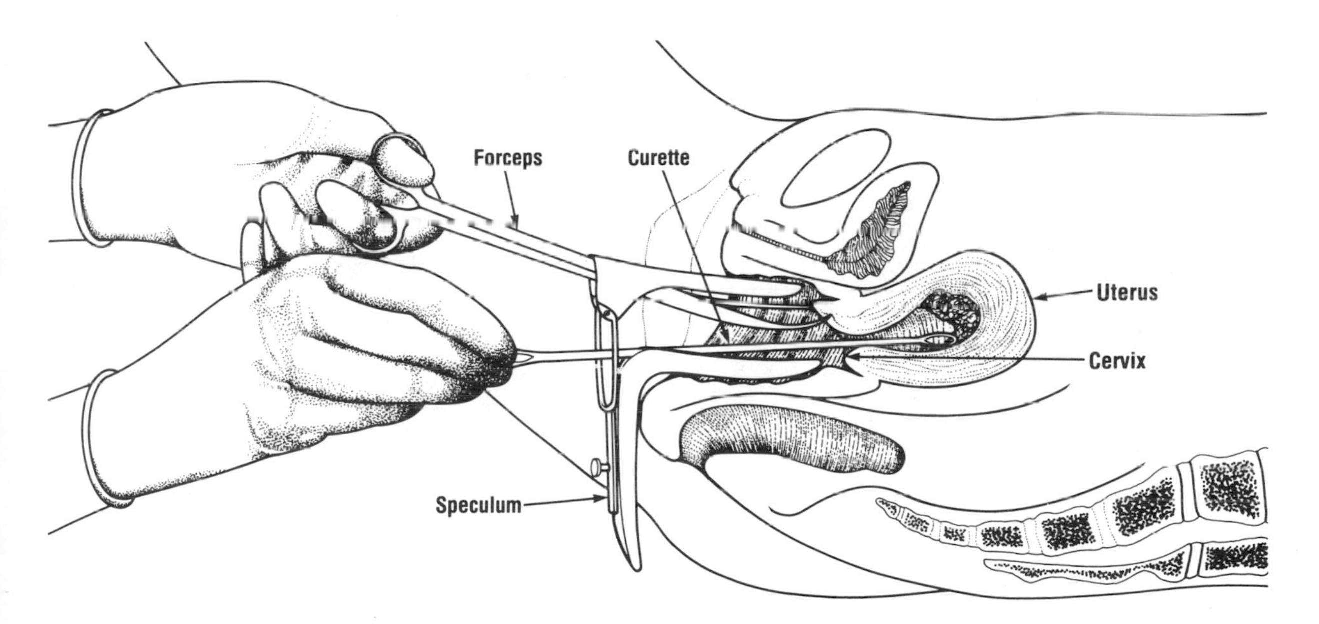

Fig. 20.8 Dilatation and curettage

3. How can a couple find out where to obtain family planning advice in your area?
4. Why is a speculum used when a woman has an internal examination?
5. Find out how the 'safe period' method of birth control using the basal body-temperature thermometer works.
6. Find out about the sheath, or condom, as

a method of contraception, using these headings:

- advantages and disadvantages of the sheath
- effectiveness
- safety in terms of health
- interruption of love-making
- sensitivity
- check-up needed
- where you can get it
- whether a prescription is needed
- problems
- general comments

7. Find out about the contraceptive pill:

- How effective is the pill?
- What is the meaning of low-dose and high-dose pills?
- What are the dangers of forgetting to take a pill?
- Why should a couple use another method of contraception if the woman on the pill has had sickness and diarrhoea?
- What are any possible side-effects of the pill?
- What diseases may be aggravated by taking the pill?
- At what age do doctors recommend a woman comes off the pill?
- What are the dangers of smoking and being overweight if on the pill?

Discuss the advantages and disadvantages of taking the pill as a method of birth control.

8. There are various forms of the IUD. Find out what these look like and draw them:

- the Copper-7 (Gravigard)
- the Copper-T
- Lippes Loop
- Progestasert

9. It is thought that, in rare cases, the IUD causes certain pelvic infections. Find out the symptoms of a pelvic infection.
10. Find out the meanings of these words used in connection with a woman's sterilisation: tubal ligation, laparoscope.

    How is the operation carried out?
11. Find out how a vasectomy is carried out.
12. The 1967 Abortion Act made it legal for abortion to be carried out. Find out about the law:

- Who introduced the bill before it became law?
- What were considered to be grounds for an abortion?

13. Think of some reasons why a woman may need an abortion.
14. Look at this chart showing the failure rates of various methods of birth control over a 12-month period:

Method	Failure rate (%)
Rhythm	25
Withdrawal	25
Sheath	5
Cap and spermicide	4
IUD	3
Pill (combined)	0.25
Nothing	70

From this chart and the information you found out for questions 6, 7, 8, 9, and 10, work out which method of birth control would best suit the following couples:

- Tom and Jenny – married with two children – are planning a third baby eventually but would like to use birth control to space out their family. They would like a fairly reliable contraceptive – but a pregnancy would not be too much of a catastrophe.
- Paula and Phil – engaged and saving for a house – are both working. They plan to have a baby in a few years' time. They need an effective method of birth control.
- Sue and Graham have one child already and are planning to have another in a couple of years or so. Sue has tried the pill and found she had a few side-effects – so they would like a method of birth control that is effective, but doesn't involve taking a drug.

15. Although abortion is legal in this country, there are many people who believe that it is wrong. In small groups, discuss abortion. Cover these points:

- Is abortion always wrong, or are there times when it is right?
- Should the man have a say in the matter of abortion?

21 Preventive health care

In order to stay as fit and healthy as possible, we need to do four things:

- eat a good diet
- take regular exercise
- avoid too much stress and anxiety
- avoid smoking, drinking, and drugs

This is not simply in order to prolong our lives, but also to improve the quality of our everyday lives. Obviously, there will be times when we are ill, but following these guidelines will help to prevent certain avoidable illnesses and keep our bodies in healthier working condition.

Healthy eating

Healthy eating is very topical at the moment. The media have been extolling the virtues of good eating habits for some time now, and programmes showing heart disease, and other results of a poor diet, have caused quite a stir. Even food manufacturers have taken note of the change in public thinking, and it is now possible to buy more nutritious foods at the local supermarkets. Basically, the four rules for healthy eating are:

- less fat, especially animal fats
- less sugar
- less salt
- more fibre

Research has shown that the richer the country, the more likely the inhabitants are to eat a high-fat and high-sugar diet in order to get their energy. A common health problem in these countries is *obesity*, or fatness. At the other end of the scale, people in the poorest countries get their energy from a diet high in carbohydrate (starch) and are likely to suffer from *malnutrition* diseases.

Sugar

In the UK, we eat a great deal of sugar. Sugar is bad for us for two reasons – it is bad for our teeth, and it makes us fat. Because sugar is easily digested, we get an immediate 'kick' from the quick release of energy after eating a bar of chocolate, say, and that makes us crave for more sweet things. Even people who try to cut down their sugar intake may find that they are still eating the 'hidden' sugars that are present in so many of the manufactured food products we buy.

Fibre

Much of the food we buy is refined and processed: the various factory processes used to manufacture the foods remove much of the fibre, such as the husks of wheat, to make the foods look more attractive to the consumer. It is now thought that lack of fibre in the diet of the richer countries has caused many health problems, and there has been much research carried out to back this up. Recommendations from the Royal College of Physicians and NACNE (the National Advisory Committee on Nutrition Education) say that we should eat more fibre than we do at present.

Salt

Too much salt has been linked with high blood pressure. Salt is added to food as it is cooked, and many people will sprinkle salt on to a meal before they have even tasted it. Yet, even if we make a positive effort to reduce the salt we add to our food, manufactured foods contain a surprisingly high level of salt.

Processed foods and additives

Nowadays, some people are concerned about the ways in which our food is processed. Refining foods removes fibres, as we know, but it also removes many of the natural vitamins, minerals, and proteins. Although manufacturers replace some of these, they do not replace them all. Another cause for concern is food additives, which are used to preserve, colour, and flavour our foods. In the past, food was preserved by salt or vinegar and flavoured by salt, herbs, and spices. Today, the list of additives is around 3000, and many of their long-term effects have not been fully researched.

People who are concerned about the food they eat should try to avoid eating processed foods, which tend to be high in fat, sugar, and salt, low in fibre and natural vitamins, and contain additives. With some planning, eating natural foods need be no more expensive than refined foods but can broaden your diet and improve your health.

Exercise

Exercise has increased in popularity over the past few years. A great many female celebrities have written exercise books and made exercise tapes in order to cash in on the current health boom. Although this may be a passing fashion, there is no doubt that exercise is good for you. A person who feels tired and lethargic at the end of a long day at work will feel refreshed and lively after an exercise session. It is a fact that exercise – coupled with a good diet – can help reduce the incidence of heart disease and high blood pressure as well as improving the efficiency of the heart muscles. Regular exercise stimulates the body to release into the bloodstream a hormone that reduces depression and aids relaxation.

Apart from this, exercise helps us to burn up more calories by speeding up the metabolic rate. This means that regular exercise will burn up body fat for up to 24 hours afterwards. People who exercise regularly find that their appetite is reduced, and they feel hungry only when they need food, rather than because they are bored or depressed. It seems that exercise makes us look and feel better and remain healthier, but it is important to start exercising gently and build up your stamina gradually. If there are any health problems, a doctor should be consulted before starting any exercise regime.

Stress and anxiety

Stress, anxiety, and tension seem very much an integral part of the modern world. It is necessary to have a certain level of these in our everyday lives to give us the momentum to get things done – too little leads to apathy and depression. Stress becomes dangerous when it goes above the level each individual can cope with, and that varies from person to person. In our personal lives, stress can be caused by work, divorce, bereavement, unemployment, retirement, health problems, pregnancy, and so on. Stress can also result from noise, pollution, and lack of sleep and exercise.

The body reacts to stress by producing the so-called 'flight or fight' hormones, *adrenalin* and *noradrenalin*, which prepare the body to cope with the stress situation. Blood will be redirected to the brain, heart, lungs, and muscles and away from other parts of the body. If you are stuck in a field with a bull, this will help you escape quickly. The hormones will be dispersed and the body will return to normal with no long-term harm done. If the stress is caused by an emotional problem with no foreseeable end, then the build-up of tension will be bad for the body as there is no means of 'running off' the tension. In time this may cause physical and mental problems. The stressed person may suffer physically from high blood pressure, digestive problems, and even heart attacks, or may develop depression, phobias, and other mental problems.

Although many people are prescribed drugs by their GPs to deal with stress, drugs only temporarily hide the stress symptoms – the underlying cause of the stress will still be there when the person stops the drug treatment. The answer lies in the person's own hands: he or she needs to take control over his or her life and plan ways of avoiding, or reducing, stressful situations. In many ways, it is simply a case of getting things into perspective – so many people worry about what might happen in the future, causing themselves unnecessary stress, rather than making positive plans for the future in order to motivate themselves in the present.

Too much work really does make Jack a dull boy. We all need time to ourselves in which to do what we want to do – either exercise, sport, or a hobby, to help us relax and recharge our batteries for a few days.

If personal relationships are causing serious stress, then professional advice may be helpful. Most relationships cause stress at some stage, as couples learn to understand each other. Honesty and open-mindedness are vital factors of a good relationship, and partners need to be ready to talk through any situation that might occur.

Relationships at work should be approached with a similar attitude. It is better to be honest with a colleague if you are feeling annoyed, rather than bottle up the bad feeling and bear resentment for weeks to come. That is not good for you or the relationship.

It obviously takes time to alter daily life patterns to achieve a less stressful day-to-day life, but many people have found the effort to be worthwhile.

Eating a good diet, exercising regularly, and reducing stress are all ways of preventing ill health from developing. This idea of preventive health care, apart from saving the National Health Service money in the long run, improves the quality of our lives. It seems far better to prevent the illness than to treat it once it has developed.

Preventive medicine and health screening

The NHS does play a part in preventive medicine. The vaccination programme aims to wipe out the diseases that were once so dangerous, such as polio, diphtheria, and tuberculosis. With the use of the rubella vaccine, it should be possible to prevent women giving birth to babies handicapped as a result of German measles during pregnancy.

Before conception

Pre-conceptual care (covered in chapter 2) is also a form of preventive medicine, as it is a way of increasing a couple's likelihood of producing a healthy baby.

During pregnancy

During pregnancy, the mother is checked regularly to ensure that no potentially harmful problems develop. The doctor and midwife will be aware of the mother's previous medical history, and will use this information to help them make any decisions about her health. A mother with a previous record of premature birth will be closely monitored to try to prevent this happening again. If the baby is born prematurely, further care will be taken to prevent infection from reaching the baby, as this is one of the greatest dangers for premature babies.

The baby

Once a baby goes home, the health visitor will give advice to parents on feeding and baby-care, in order to ensure that the baby is as healthy as possible.

Within the first couple of months, the baby will be tested for phenylketonuria (PKU), a very rare metabolic disease which can lead to mental retardation. If found, this disease can be treated and prevented. Special screening is carried out if there is reason to believe that the baby is at risk of developing a handicap as a result of:

- family history, as many handicaps can be inherited
- an abnormal pregnancy, e.g. if the woman developed rubella or toxaemia
- problems after the birth, such as the baby's contracting a serious illness or having convulsions

The baby clinic and health visitor take care to check that the healthy baby develops normally.

Infectious diseases

In later life, many *communicable* – that is, contagious or infectious – diseases can be prevented by immunisation, and many potentially fatal diseases have been reduced since the immunisation programme became widespread.

Apart from immunisation, there are other ways of preventing the spread of highly infectious and dangerous diseases. *Epidemiology* is the study of how diseases are spread. The intention is to find out how and try to prevent more cases of the disease.

Many infectious diseases are *notifiable*, meaning that they must be reported to the Environmental Health Officer. This will show where various infections have occurred and will help towards an understanding of the causes of the disease.

A detailed medical history of the patient will be taken in order to see if there is a link that might show the source of the infection. In many cases, diseases are spread by a carrier who may personally show no symptoms of the disease.

Screening tests

There are various screening tests available on the National Health Service and in private health clinics. Screening tests aim to detect illnesses or conditions that have not yet become apparent, and they are a very necessary part of preventive medicine. They are covered in more detail in question 23 of the assignments.

Assignments

1. What have you eaten recently? Write down everything you ate yesterday – and be honest about it. Was it a good diet? Bearing in mind 'less fat, sugar, and salt and more fibre', are there any improvements you should be

making?

2. Healthy eating is not simply a case of not getting fat – it is also necessary to keep the internal body system in good working order.
   - What damage does too much fat in our diet, especially animal fat, do to our bodies? Check that you understand the meaning of polyunsaturated and saturated fat.
   - What damage does too much sugar in our diet do to us?
   - What damage does too much salt do?
   - What damage does too little fibre do?
3. Make a set of guidelines to help reduce the amount of fat we eat in our everyday diets.
4. Arrange to visit a local supermarket and look at some of the food labels for items such as: tomato ketchup, pickle, baked beans, fruit yoghurts, salad dressing, tinned spaghetti, tinned vegetables, muesli. These all contain sugar. Why do you think this is?

   Labels have to show ingredients listed in order from the highest proportion downwards. Where does sugar come on these labels?
5. Make a set of guidelines to help someone cut down on sugar intake. (Artificial sweeteners are not a long-term solution, as they do not help to reduce a sweet tooth.)
6. How can we get the maximum fibre content when buying or preparing these foods: rice, flour, bread, pasta, potatoes, apples?
7. It is thought to be better for us to eat more vegetable protein and less meat protein, as meat is high in fat. Find out some sources of vegetable protein.
8. Visit the local supermarket again – this time to look at labels for salt (often called sodium). Look at bacon, savoury biscuits, corn flakes, ready-made meals, crisps, sausages, smoked fish, cottage cheese, and tinned ham.

   How high up on the list of ingredients is salt in each case?
9. Collect these leaflets from the Health Education Council: 'Fibre in your diet', 'Fat, who needs it?', 'Eating for a healthy heart', 'Food for thought', etc. As a group, discuss the set of leaflets. Include these points in your discussion:
   - presentation
   - information
   - target audiences
10. There is at present much controversy over food additives. Obtain a copy of 'Food additives' from FDK, 25 Victoria Street, London SW1H OEX, or *E for Additives* by Maurice Hanssen, published by Thorsons, Wellingborough, Northants.

    a) Use either or both of these publications to help you understand the uses of food additives:
       - Preservatives and stabilisers – what are their functions?
       - Colourings and flavourings – what are their functions?

    Do you think that one of these two areas of food additives is more important than the other? If so, why?

    Research has shown that several food colourings used in food manufacture may not be as safe as previously thought. In fact, baby foods are no longer allowed to contain any colourings.

    b) Some evidence suggests that hyperactivity in children may be partly caused by food additives, particularly E102 (Tartrazine).
       - Find out about the signs and symptoms of hyperactivity.
       - Find out about E102 (Tartrazine).

    c) As a group, discuss your views and opinions on food additives. Over the next few days, look at some food labels and see just how many additives you eat.
11. Check your pulse-rate for fitness:
    - Take your resting pulse (the pulse-rate when you have done nothing vigorous) and count the beats in one minute.
    - Take your pulse during exercise. Run on the spot for one minute and take your pulse for one minute.

    To keep to a healthy pulse-rate, subtract your age from 200 (e.g. $200 - 18 = 192$ for an 18-year-old), and do not go above 75% of this figure (e.g. $\frac{200 - 18}{100} \times 0.75 = 144$ beats per minute for an 18-year-old).
12. For the group, work out an exercise routine that works on the
    - neck, arms, and hands

- chest and shoulders
- abdominal muscles
- back
- waist
- hips and bottom
- legs

Remember to warm up and cool down in the exercise. Why is it particularly important to warm up before starting exercise?

13. What is the difference between aerobic and anaerobic exercise? What is the particular value of aerobic exercise?
14. Look at the publication 'Look after yourself', available from the Health Education Council, to help you work out the value of the following sports: cycling, aerobic work-out, weight-training, jogging, swimming, squash, martial arts.
15. Exercise can be particularly useful for keeping fit in old age and for helping arthritis sufferers. How do you think exercise helps these people?
16. What do you think 'flight or fight' means when describing the hormones adrenalin and noradrenalin?
17. What do you think are some of the physical and mental signs of stress?
18. Write a set of steps aimed at reducing stress in our everyday lives. Here are some suggested areas to cover:

- work
- leisure
- relationships (at work, family, and sexual)
- health
- eating patterns
- emotions
- self-esteem

19. There are various ways of relaxing open to us – some are simple and some more complex. Find out briefly about some of the following:

- yoga
- meditation
- massage, and general body and facial treatment at a beauty parlour
- jogging
- hypnosis
- reflexology
- biofeedback
- autogenics

20. Find out about vaccines for the following:

- rubella
- tetanus
- rabies
- anthrax
- typhoid
- yellow fever
- cholera

Find out when they are given and whether they need to be repeated.

21. Find out which are the notifiable diseases.
22. Some diseases can be prevented. Briefly, how could these infections be avoided:

- food poisoning
- rabies
- STDs

23. Find out more about these screening tests:

- mass X-ray
- cervical smear
- breast examination
- blood test
- sight test
- hearing test
- medical check-up
- dental check-up

- What can they detect?
- Explain what happens.
- Who is eligible for them?
- What is the cost, if relevant?

22 Safety in and around the home

If we can become more aware of the possible safety hazards in our everyday lives, then we will be one step nearer to reducing the high number of injuries and deaths caused by accidents each year.

Safety at work is covered by the Health and Safety at Work Act 1974, and there are many laws regarding road safety, storage of dangerous substances, safety standards in electrical and gas equipment and appliances, and so on. However, accidents still happen, because many people are unaware of the potential dangers. The home and its surroundings are an accident black-spot, and there is no Health and Safety at Home Act to set guidelines for us, so we need to look at the everyday dangers at home and in the garden.

About 6000 people die each year as a result of accidents in the home, and many more are injured. Babies, toddlers, and the elderly are most at risk: babies and toddlers because they have not learnt caution from experience, and the elderly because their senses may have deteriorated and they are often frail and infirm.

Babies

The danger to look out for with young babies is choking – babies should never be left with a bottle, or near small items that they could put in their mouths.

Toddlers

Toddlers are very inquisitive and explore everything they can in order to learn, so poisonous household cleaners need to be stored safely, and thought must be given to the dangers of electrocution. Toddlers are not very steady on their feet and can easily fall into glass doors or unguarded fires.

Adults

In adulthood, many accidents are caused by human error. Electrocution is quite common, as untrained people attempt repair and maintenance work on electrical appliances. Home decorating, where ladders or trestles are used, is another hazard, as falls, even from a small height, can cause serious injuries.

The elderly

The elderly may suffer from arthritis or rheumatism, which makes them more likely to fall and hurt themselves. Failing sight means they may miss many possible dangers, and forgetfulness causes other problems – they may forget that they have left the gas on, for instance.

In the house

The kitchen

The kitchen is one of the most dangerous places in the house (fig. 22.1). The hazards of electricity, gas, poisonous household cleaners, and heat are all together in one room.

Cooking is one of the most hazardous activities carried out in the kitchen. The cooker can burn children, especially the newer type of electric cooker that does not show a radiant heat. Chip pans cause many fires if they are overfilled with fat or left unattended. Touching electrical switches with wet hands can lead to electrocution, as water is a good conductor of electricity, and children pulling the flex of an electric kettle leads to cases of scalding.

Parents of young children should be particularly careful about keeping household cleaners well out of reach – the bottles tend to be brightly coloured and attract children's eyes. Polythene bags and carrier bags also need to be stored out of reach to prevent the danger of suffocation. The elderly, and anyone else with poor eyesight, need to take particular care in the kitchen.

The living rooms

The remainder of accidents which happen downstairs in the sitting-room and dining-room areas are mainly caused by fires (fig. 22.2). Mirrors and

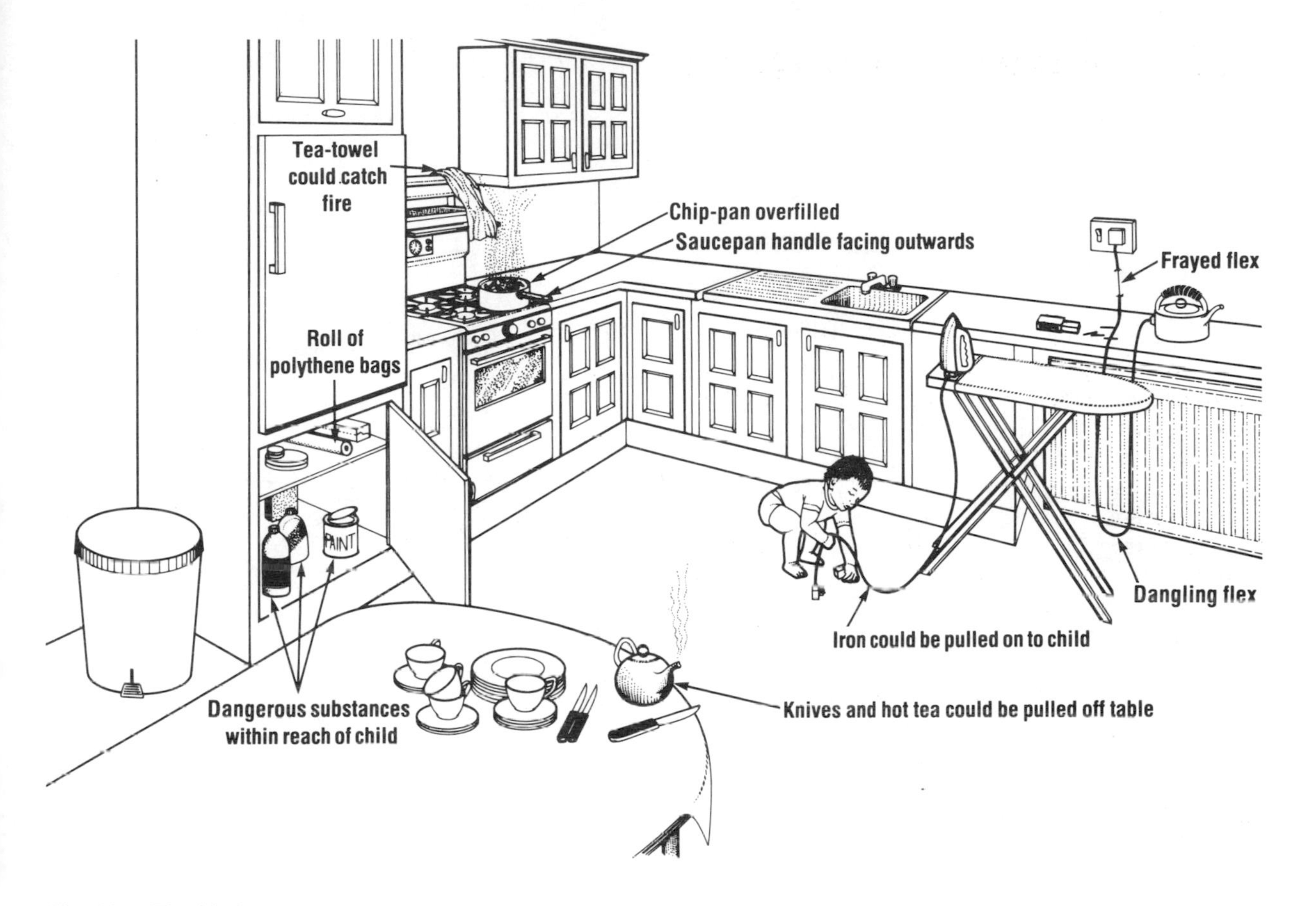

Fig. 22.1 The kitchen

shelves should not be put above an open fire as clothes can catch fire very easily, and television sets should always be unplugged when not in use. There have been quite a few instances recently of fires caused by furniture upholstered with polyurethane foam, so it is a good idea to check that your sofa and chairs are safe.

The stairs

The stairway (fig. 22.3) can be dangerous for everybody but, again, young children and the elderly are most at risk as they are unsteady on their feet. It is possible to reduce the risks by making sure that the stairs are uncluttered by toys, etc., and having them well lit. A stair-gate is a useful way of preventing toddlers and babies from going up or down stairs. To be extra safe, babies should be shown how to go down stairs as soon as possible.

Many newer houses have ladder-like stair treads and balustrades, which are dangerous as young children may try to climb up through them. Radiators, tables, or glass doors at the bottom of the stairway are likely to add to the injuries of anyone unlucky enough to fall down the stairs.

The bedroom

The bedroom holds a surprising number of safety hazards (fig. 22.4). Fires can be caused by a faulty electric blanket, or by falling asleep while smoking in bed. Hot-water bottles can burst and scald, so check that the rubber is not perished. Babies should never be put to bed with a feeding bottle, as they could choke. Pillows are dangerous for any child under the age of one as they can cause suffocation.

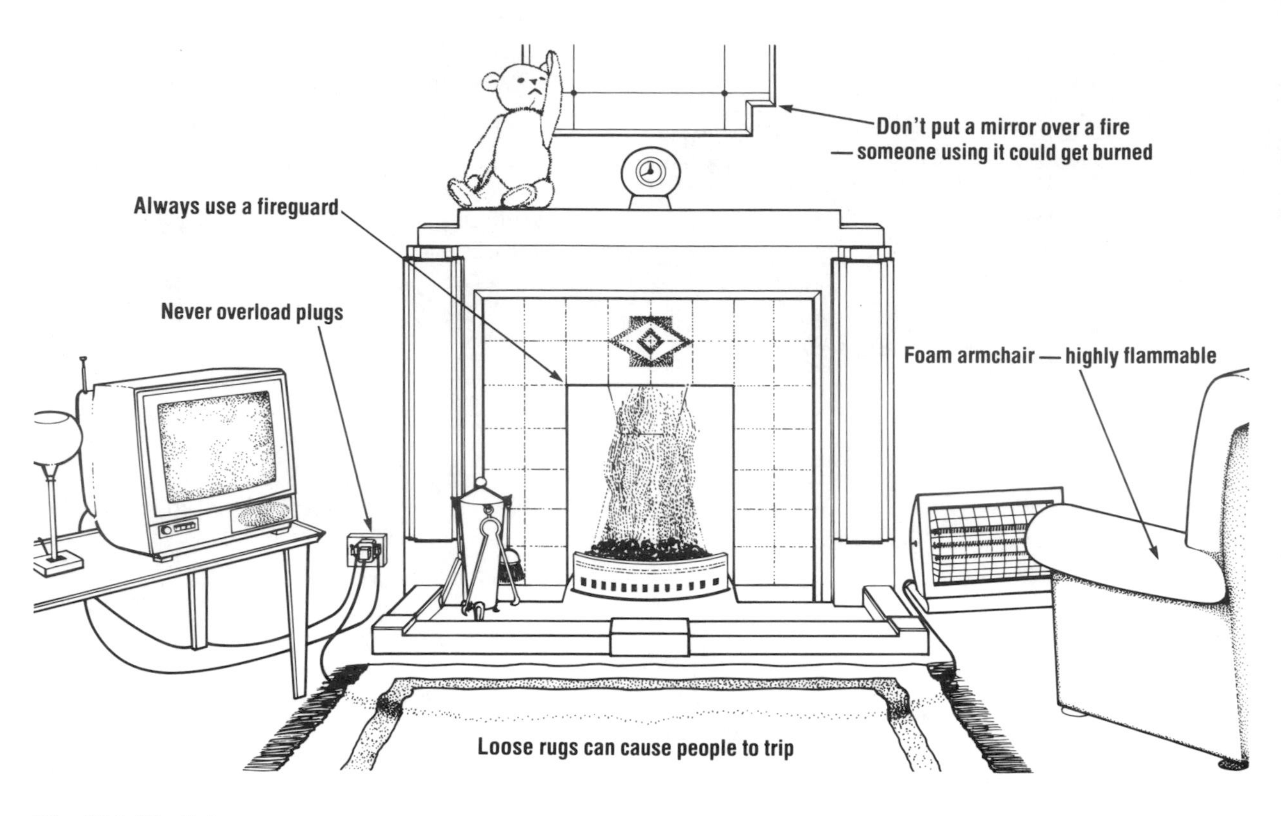

Fig. 22.2 The living room

Bathroom

The bathroom, rather like the kitchen, has the hazards of water, electricity, heat, and occasionally poisonous substances such as drugs, all together in one small room.

It is a legal requirement that light switches in the bathroom must be on a pull-string to avoid contact with wet hands, and portable electric fires should never be brought in to heat up a cold bathroom. Any electric bathroom heaters need to be specially fitted on to the wall, and should have a pull-cord.

Babies and children need special attention at bath-time as they could be scalded, or drown. The cold water should be put in first and hot water be added to bring the bath up to the right temperature. Babies should never be left unattended in the bath, even for a few minutes.

Non-slip rubber mats that fit inside the bath are useful to prevent the old and young from slipping over, and hand-rails offer something to hold on to. Many people keep their medicines in the bathroom, so these should be stored well out of reach. Any out-of-date or unfinished medicines should be either flushed down the toilet or taken back to the chemist. Children, being curious, are quite likely to play with the lock and lock themselves in the bathroom, so locks should be fitted high up and out of reach.

Outside the house

There are other hazards outside the house. The garden contains poisonous substances. Some occur naturally, such as poisonous leaves and berries, and others are found in the many poisonous products stored in the garden shed, such as weed-killer, paint-stripper, etc.

With the use of electrically powered lawn-mowers, there has been an increase in the number of accidents in the garden, so people should be particularly careful to follow the manufacturers' instructions.

Fig. 22.3 The stairs

Babies and toddlers can drown in a few inches of water, so ornamental ponds should be made child-proof or be filled in.

Assignments on safety in and around the home

1. Answer these questions on safety:
   - Why are young children and the elderly more likely to have an accident?
   - How could household cleaners be safely stored?
   - How can electric sockets be made safe from prying fingers?
   - How can glass doors be made safer if someone should fall into them?
   - How might poor eyesight and forgetfulness lead to an accident for an elderly person?
   - How can a cooker be made safer for young children?
   - How should you deal with a chip-pan fire?
   - Why should cold water be added to a baby's bath before hot?
   - How have medicine containers been made safer?
2. Discuss the possible dangers in each of these areas for a toddler and an old person: kitchen, stairs, bathroom, bedroom, lounge/dining-room. List all the safety hazards there are in each area, and decide what safety measures you could take to help prevent an accident.
3. Fire can happen anywhere in the house, but there are steps we can take to help reduce the risk of fire. Using these ideas, design a fire-prevention checklist:
   - matches and cigarettes
   - open fires
   - electrical appliances and wiring
   - gas appliances
   - paraffin
4. Various symbols on the goods we buy tell us that the product has been tested for safety. Find out what these symbols are.
5. Many products we buy have safety warnings printed on them. Find as many of these warning labels as you can. Either copy them down or collect examples of them. Note whether the warnings are: (a) easily seen, (b) written in simple English.
6. Find out which leaves and berries are poisonous. What symptoms do they cause when eaten?
7. What sort of accidents are caused by lawnmowers? How could these accidents be avoided?
8. If a baby is to be left unattended in a pram out in the garden, what precautions should be taken?
9. Obtain some copies of the leaflet 'Play it safe', published by the Health Education Council. It covers all aspects of child safety and gives simple, easy-to-follow advice on first aid. As a group, discuss the leaflet. Include these points in your discussion:
   - Presentation. Do you like the way the first-aid advice is included alongside details of the safety hazards?

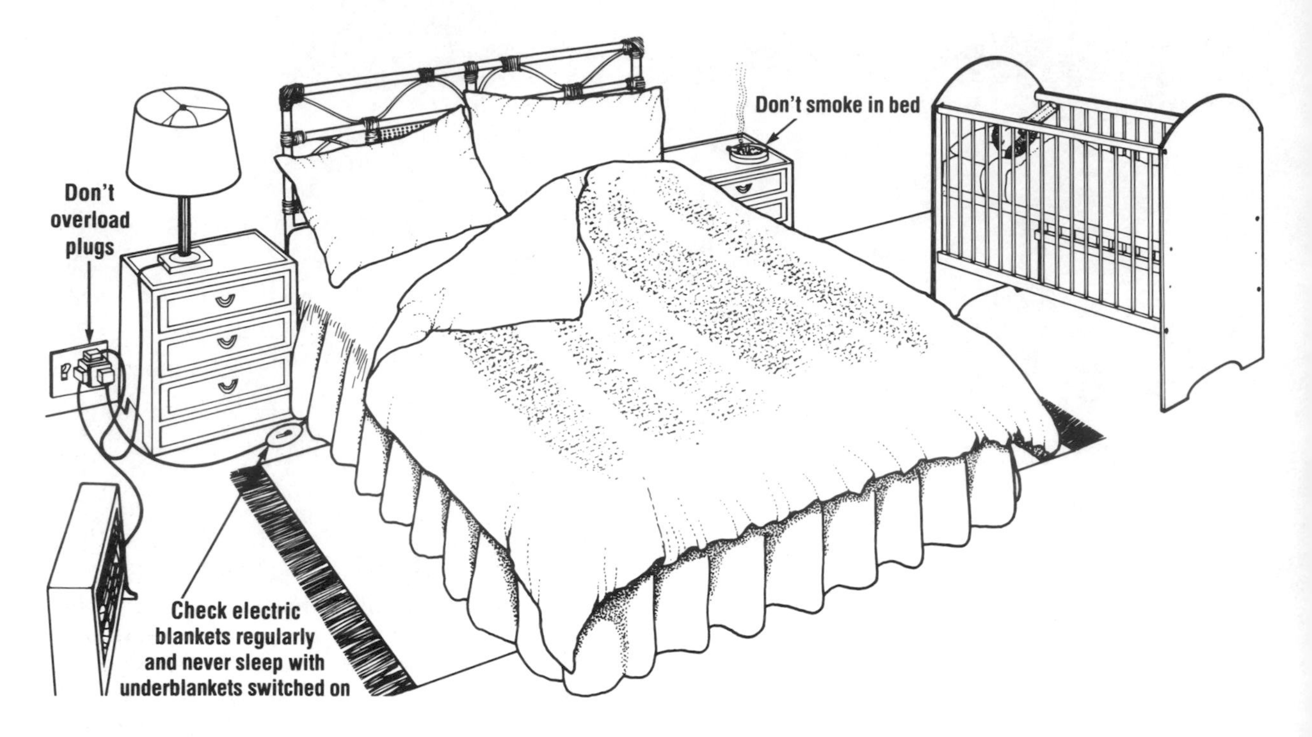

Fig. 22.4 The bedroom

- Are the first-aid instructions clear and easy to follow? Could someone with no knowledge of first aid understand them?

Carry out some of the first-aid instructions yourselves.

10. Find out the prices of these pieces of safety equipment: stair-gate, cooker guard, car safety seat, safety socket, inflatable armbands for swimming, child-proof locks for kitchen cupboards, fireguard, window locks.
11. What are the possible dangers in these situations:
    - leaving the car engine running in the garage
    - using adhesive in an enclosed space
    - parking close to the school gates
    - wearing a scarf on a moped or motorbike
    - mixing household bleach with other cleaners
    - putting a cooker near a window

23 Talking to your doctor

Visiting the doctor

Most of us are registered with a doctor – a general practitioner – even if we do not visit him or her very often.

Although patients usually visit the doctor because they are unwell, doctors are also involved in preventive medicine – i.e. making sure we stay healthy. Examples of this are the vaccination programme, which is intended to keep us free of disease, cervical smears for women, medical examinations, and so on.

Many people feel nervous when they go to the doctor's. There are a number of reasons for this:

1. They are worried about their health and are seeking professional guidance from the doctor to put their minds at ease, and this fear may cause them to 'clam up'.
2. They may feel ill-at-ease with the doctor – either because they see doctors as 'distant professionals' or because they have not yet built up a relationship with the GP.
3. They may lack the confidence to explain their symptoms to the doctor. Many people find it difficult to communicate and get very worried when faced with a situation such as a visit to the doctor's.

In the past, patients' attitudes to doctors were very respectful, and many doctors felt entitled to be treated in this way. More recently, people began to question the established order of things in society, and this questioning also affected the doctor/patient relationship. Doctors were no longer raised on a pedestal but were seen as professionals trained to do a job, as well as human beings.

Nowadays, our relationship with the doctor is changing, and most doctors and patients would say that this change is for the better and that attitudes are improving in doctor/patient relationships.

However, it is important to remember that doctors need to keep their relationships with patients on a professional footing so that they can make objective decisions. It is important to remember, too, that respect is a two-way process: the doctor must also respect the patients' needs.

In the past, a person with a problem could usually find someone to discuss it with. There was the support of the church: the vicar was there to help in times of need, and the Bible provided a good basis on which people could model their lives.

There was also the support of the *extended family*, where uncles, aunts, grannies, and grandads lived either together or very close by. Now we have what is called the *nuclear family*, where the family unit is small – just parents and their children – and frequently living miles away from any close relatives.

In the past, people could talk about any worries with their family before they became real problems. Without this type of support – the church has become less important and families have become more insular – people now need to take their problems to the doctor, as there is no one else to listen. The doctor has to deal with the total health of patients, not just their medical problems, and this puts a great burden on him or her. This burden could be reduced if the services of a professional counsellor were offered to patients who need time to talk about their worries. In some practices this is already happening.

Many patients worry about explaining their symptoms to the doctor, and fear that they might forget something important. Although making a written list may encourage mild hypochondria, it is a good idea for the patient to consider these questions:

- Why exactly am I going to the doctor's?
- What do I hope to get from the visit (e.g. reassurance, prescription)?

Asking these questions will help to make the symptoms clearer in the patient's mind, and he or she will be more likely to remember them once in the surgery.

Some patients may go into the doctor's surgery having already made a diagnosis and thinking they know what medicines they want. Again, it is important to remember that a doctor is only human, and this type of approach is likely to annoy and antagonise even the most patient and caring GP, as well as undermining his or her authority in the medical field.

However, a doctor is likely to be equally irritated by patients who have not given their symptoms any thought and expect the doctor to make a diagnosis without knowing any details.

Recently, there has been a great deal of coverage, both in the press and on television, of how patients are often kept in the dark by the medical profession. This type of criticism can be constructive, as it encourages the medical profession to take stock of itself and to improve any areas found wanting. These changes have made the medical profession more answerable to criticisms, and more aware of the needs of the consumers – their patients. The community health councils exist in order to help us cope with any problems and queries we may have on any aspect of the National Health Service.

Although attempts are being made to consider the needs of the consumer in the health service, some consideration also needs to be given to ways of helping the patient to help himself or herself.

Professionals working in the health-service and health-education fields are shocked by the level of ignorance about the way our bodies work and the firm beliefs in many of the old wives' tales that have been around for years. Sadly, even with the improved educational opportunities available today, this ignorance does not appear to be lessening. This lack of understanding also affects the patient's ability to interpret the doctor's diagnosis. Doctors who are aware of this problem try to explain the diagnosis and treatment in simple terms, but the patient may still misunderstand much of what is being said.

These problems suggest a need for more health education in school, starting at a younger age. It is important for everyone to understand how the body works.

The pharmacist

At every dispensing chemist, which is where prescriptions are made up, there has to be a qualified pharmacist. The pharmacist's job is to ensure that the medications desired are the ones prescribed by the doctor, and that they are made up to the correct strength.

As well as writing out the dosage instructions for the medicine or pills, the pharmacist will explain to the customer what to do, and answer any questions.

People will sometimes ask the pharmacist's advice about a minor ailment, rather than bother the doctor with something they are concerned about. The pharmacist can then suggest medications that can be bought over the counter. He or she is not allowed to sell those drugs which are available on prescription only.

For minor illnesses, such as colds and coughs, it is not really necessary to go to the doctor's unless the illness becomes worse rather than better or the patient is very young, very old, or has other health problems. If patients can go to the pharmacist for advice with these minor ailments, then the doctor will have more time available for needier cases. It is reassuring to know that the pharmacist would also refer patients to the doctor, if necessary.

In general, we do not make enough use of the skills and knowledge of our pharmacists, mainly because we do not realise what services they offer, and also because some people feel that only the doctor knows best. Perhaps this last view will have to change quite soon, as doctors' work-loads will become heavier when the old and infirm are returned to the care of the community, rather than treated in hospitals and other institutions.

Taking the medicine

Depending on the doctor, anything from 30 to 60% of patients leave the surgery with a prescription for some form of medication. Unfortunately, many patients do not feel satisfied if they leave their doctor's surgery without a prescription, and it will take some time to change attitudes about this.

However, of the medicines that are dispensed to the patient, it is a sad fact that many people do not understand the instructions printed on the label and, more frequently, do not understand what the medicine is for. Although the pharmacist is there to help and advise the patient, many patients forget their instructions as soon as they get home. Unfortunately, not to follow the instructions could mean that the medicine is virtually useless, or may cause unwanted side-effects (covered in the assignments).

Assignments on doctors and pharmacists

1. Find out about your own doctor's surgery. Here are some ideas, but do add more questions yourself:
   - How many doctors are there in the practice?
   - Apart from the doctor(s), what other staff are there?
   - Which of these additional services are available:
     - pre-conceptual advice (FPA)
     - ante-natal clinic
     - baby clinic
     - counselling
     - women's clinic
     - social worker
     - other (please state)
   - What are the surgery hours?
   - Are appointments necessary?
   - If so, how far ahead do you need to book?
   - Are emergency home visits made if necessary?
   - Is the waiting-room comfortable?
   - Are children catered for with toys, books, etc?

   These questions are not intended to be critical, but to show the wide variations between practices. Comparing findings between the group, can you form any conclusions?
2. At what age can a person choose his or her own doctor?
3. As a group, discuss how a doctor might reply to these patients' comments:
   - 'Well, doctor, I've got these spots round my middle. I think it's shingles. My neighbour said if the spots meet round the middle I'll die! What can I do?'
   - 'My baby's got a temperature of 102°, but I've wrapped her up nice and warm so she can sweat it out.'
   - 'Actually, I've not had a bath for five days. I've had a period, you see, and bathing is bad for you during a period.'
   - 'My little boy keeps touching himself – the other day he did it in the supermarket! It's not that I mind or anything, but I tell him he'll have bad eyesight when he's older.'
   - 'I've got piles, doctor. I know I have, because the other day I sat on a radiator because I was so cold.'
   - 'My baby's not putting on weight as fast as the baby next door – I think my milk is too thin. My neighbour said she put her baby on the bottle because her milk was too thin.'
4. In groups of two, act out the following situations using role-play:
   a) An understanding doctor. Patient very worried. Complains about a minor ailment, a sore throat, but is really more concerned about a lump she has found in her breast – is unable to communicate this easily to her doctor.
   b) A 'traditional' type of doctor who sees himself as being in a position of power and likes respect from patients – does not like patients to be too knowledgeable about medical matters. Patient younger than the doctor. Does not need just a prescription, but would like her ailments to be explained for reassurance.
   c) An understanding doctor. Patient believes himself or herself to be as knowledgeable as the doctor. He or she has diagnosed the ailment and decided what treatment should be used – has come to tell the doctor to write the prescription.
5. As a group, discuss why professional people, such as doctors, need to keep a certain distance between themselves and their clients.
6. How important do you think health education is in schools?
7. Check that you understand the meaning of these terms: suppository, cream, ointment, lotion, pessary, tablet, pill, capsule, linctus, inhalant.
8. Arrange a visit to your local pharmacist to find out the most commonly used instructions and warnings on medicine labels. Ask him or her to explain any that you don't understand. List these instructions, together with an explanation, in a way that could help patients to take their medicines correctly.
9. Why should you not drive, or operate machinery, after taking certain medicines?
10. Why is it important to tell the doctor about any other medicines you are taking before he or she writes a prescription?
11. Why is it necessary to complete the entire course of an antibiotic?

24 Home nursing

In talking about home nursing, we are dealing with adult patients (nursing the sick child is covered in chapter 12).

The sick adult may have a minor ailment or something more serious – and this needs to be taken into account by the carer. The main point to remember in caring for any patient is that the patient is dependent on the carer to make the surroundings as comfortable as possible and to cater for the patient's needs.

The sick-room

The sick-room (fig. 24.1) should be arranged so as to make nursing the patient as easy as possible. It should be warm, but well ventilated. An ideal temperature is about 15°C (60°F), or about 18°C (65°F) for the elderly and babies.

In most cases, the patient's own bedroom is best, as this is familiar, but sometimes, as in the case of a long illness, it may be more convenient to move to a bedroom nearer to the bathroom. It is a good idea to have some means of contacting the carer, such as a bell that can be rung in an emergency.

Hygiene is important in the sick-room, so the bedding should be easy to wash and dry. If the patient is incontinent, then the mattress should be protected by a waterproof sheet. The patient's clothing needs to be warm, but light and easily washed.

Fig. 24.1 The sick-room

Diet

Diet is particularly important for the invalid. As he or she may not have much of an appetite, the food should be appetising and attractively presented. As well as needing food to provide energy, body heat, and growth, the invalid also needs food to repair the damage caused by the illness. Sometimes the invalid needs a special diet because of the illness – such as sugar-free for diabetics, or low-salt for high blood pressure – and it is essential that the dietician's orders are followed to help the patient recover.

The patient's emotional needs

Throughout the time the carer is nursing an invalid, the patient's needs should take priority. Patients have the same needs as healthy people, plus the special needs of an invalid. The carer may at times feel impatient and intolerant towards the patient, who may be irrational and miserable, but these feelings must not show. Patients, like healthy people, need to be given love and affection.

Hospital convalescent patients

Nowadays, hospitals are tending to send patients home earlier than in the past. These patients will be treated at home by the GP and the community nurse. Some may need to return to hospital for treatment as out-patients; others may attend day centres on a regular basis. The patient will still need the same care as the patient who has been ill at home from the start, but the carer will need to ensure that the hospital's instructions are carried out.

Convalescence

Once patients begin to make a definite recovery, they become convalescent. Convalescence needs to be started gradually, with more activities introduced as the patient becomes stronger.

Assignments on home nursing

1. Arrange for a nurse, or someone from the Red Cross or St John Ambulance Brigade, to come and demonstrate these aspects of home nursing:

   a) Bed-making:

   - making a bed when the patient is in bed and not in bed
   - using a draw-sheet
   - changing a top sheet

   b) The patient needs to be comfortable in bed, so particular attention should be paid to personal hygiene:

   - preventing pressure sores
   - treating pressure sores
   - giving a bed bath
   - care of hair, teeth and mouth, and nails

   c) The patient will need to be moved. The patient's comfort must be considered, and the lifter needs to lift properly to prevent back injury:

   - lifting a patient from a flat position
   - lifting a patient who is sitting up
   - turning a patient in bed
   - helping a patient from bed to chair
   - helping a patient sit up
   - helping a patient up from a chair

   (These may be carried out alone or with a helper.)

   d) Medical attention needs to be given carefully and considerately:

   - taking the temperature – mouth and under arm
   - taking the pulse
   - giving medicines, pills, and liquids
   - giving eye, ear, and nose drops

2. It is important that medicines are stored and disposed of carefully. Explain:

   - where drugs should be kept
   - why drugs should never be left on the patient's bedside table
   - why drugs should not be used after the expiry date
   - how drugs should be disposed of

3. Write brief notes on the causes and treatment of pressure sores.
4. What equipment do you need for giving a bed bath?
5. Sometimes, patients are put on a special diet. Find out about these special diets:

   - diabetic diet
   - low-calorie diet
   - low-salt diet

- low-fat diet
- gluten-free diet

6. The meal tray should be made as attractive as possible. How can this be done?
7. What needs does the patient have? Perhaps these headings may help:

- emotional needs
- physical comfort
- companionship

8. Think of some interesting convalescent activities and pastimes, both indoors and outdoors, that would help patients recover their health, strength, and interest in life. Plan activities for: (a) a convalescent child, (b) a convalescent old person.

25 Basic first aid

You never know when you might be called upon to carry out some first aid. It may be simply a case of putting a plaster on a cut finger, or it may be saving someone's life at a road accident. However, an opinion poll carried out by MORI for the St John Ambulance Brigade showed that most of the population are likely to do more harm than good in an emergency.

The main aims of first-aid treatment are:

- to save someone's life
- to prevent the person's condition from deteriorating
- to aid healing and recovery

In an emergency, quick thinking is essential. In the home, car, and at work, a first-aid kit (fig. 25.1) should be within easy reach. These items will be useful in a first-aid kit:

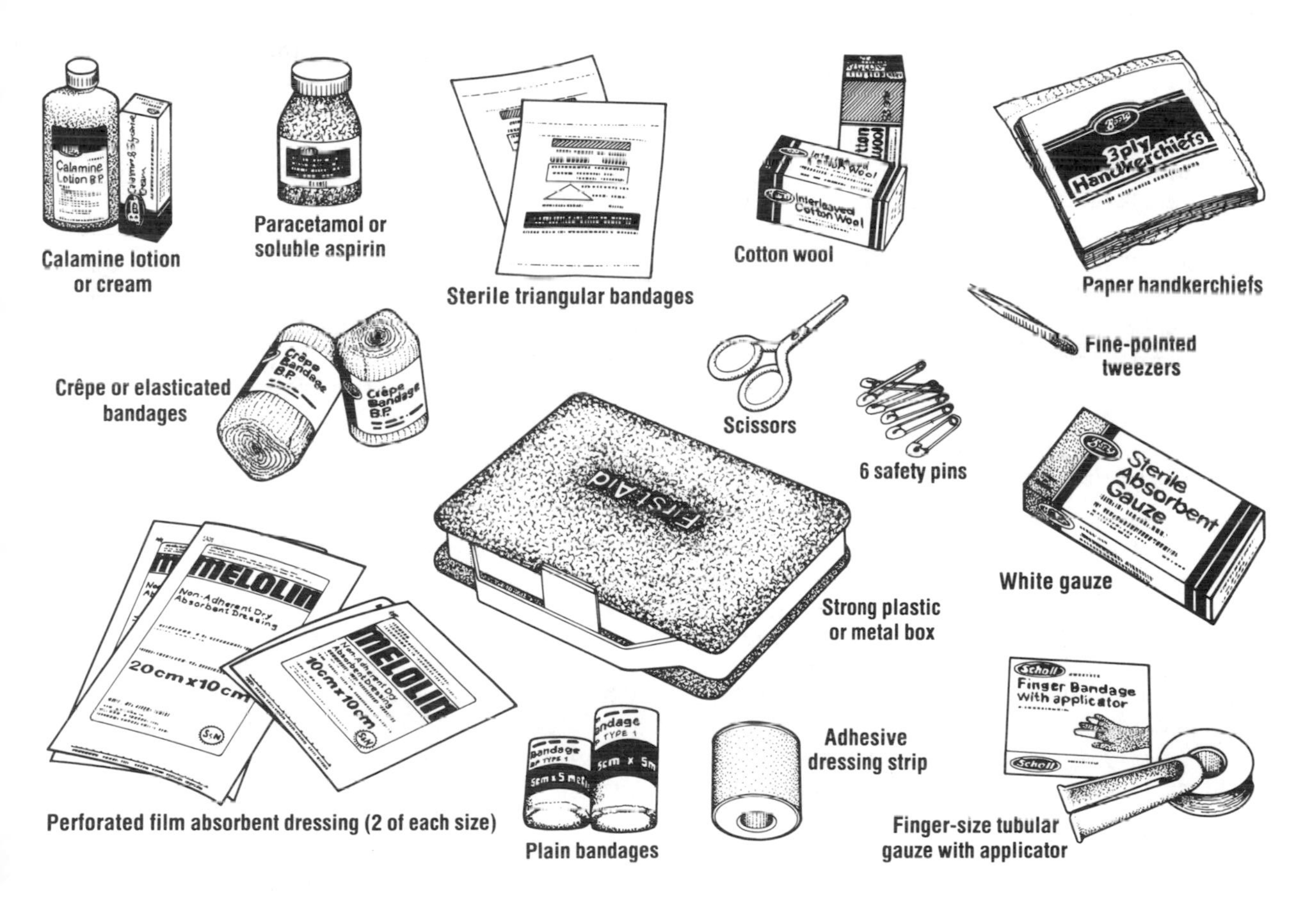

Fig. 25.1 Some items found in a first-aid kit

- an up-to-date first-aid book
- assorted sterile gauze bandages
- a small bag of cotton wool
- a triangular bandage
- Melolin non-stick dressing, 10 cm × 10 cm and 5 cm × 5 cm
- elasticated bandages, 15 cm × 10 cm
- crêpe bandages
- plasters
- plaster strip
- tubular gauze and applicator for fingers
- safety pins
- tweezers
- scissors
- a thermometer
- antiseptic cream
- pain-killers
- a 5 ml spoon
- a medicine glass
- an eye bath
- calamine

The statutory first-aid box at work has far fewer contents – but at home there are no laws governing first-aid boxes. Although the first-aid box should be easily accessible, medicines should be kept out of reach of children.

First aid is best learnt by attending courses run by the Red Cross and the St John Ambulance Brigade, but there are some procedures which everyone should be able to carry out:

- stopping bleeding
- using bandages
- dealing with the unconscious casualty
- treating shock
- dealing with choking
- getting help

Resuscitation and external chest compression are also essential first-aid techniques, and knowledge of them could help save a life. However, as they need to be learned from a qualified instructor, they cannot be included in this book. If you are interested in learning these techniques, contact your local British Red Cross Society or St John Ambulance Brigade.

When carrying out any first aid at the scene of an accident, it is important to check that you:

1. send for help from emergency services immediately;
2. only move the casualty if in any danger, e.g. from fire, toxic gases, etc.;
3. give treatment that is immediately necessary, e.g. check that the casualty is breathing, stop bleeding
4. understand the treatment of shock, and do not give any food or drink.

Shock

A casualty will often be suffering from shock, so it is essential to know how to deal with this. Shock can be caused by an accident, severe pain, blood loss, or a heart attack. This type of shock is called *clinical* or *traumatic* shock, and is not the same as the emotional shock resulting from a bad experience, or injury, where the person recovers quite quickly. Clinical shock is brought about by the blood being redirected to the vital organs (brain, heart, etc.) away from the rest of the body. Shock develops gradually and may lead to unconsciousness and death.

The symptoms of shock are:

- fast and shallow breathing, and increased pulse-rate
- skin greyish in colour, and cold and clammy to touch
- dizziness or fainting; nausea and vomiting
- thirst
- anxiety and restlessness

The way to treat shock is to lay the casualty down and raise the legs (unless injuries make this impossible) in order to help the blood reach the brain. Any tight clothing must be loosened, and the casualty should be covered with a warm, but light, coat or blanket.

Points to remember:

- Talk to the casualty and reassure him or her.
- Don't give the casualty anything to drink or eat, as an anaesthetic may be needed at the hospital.
- Don't let the casualty smoke.
- If the casualty stops breathing, start mouth-to-mouth resuscitation if you have learned how.
- If the casualty loses consciousness, or feels sick, put him or her into the recovery position, if injuries allow.

The recovery position

A casualty who is unconscious but who is breathing and whose heart is beating should be put in the recovery position (fig. 25.2). This position is good for three reasons:

1. The airway is kept open to allow the casualty to breathe.
2. The tongue cannot drop down the throat.
3. If the casualty vomits, it will come out of the mouth and not cause choking.

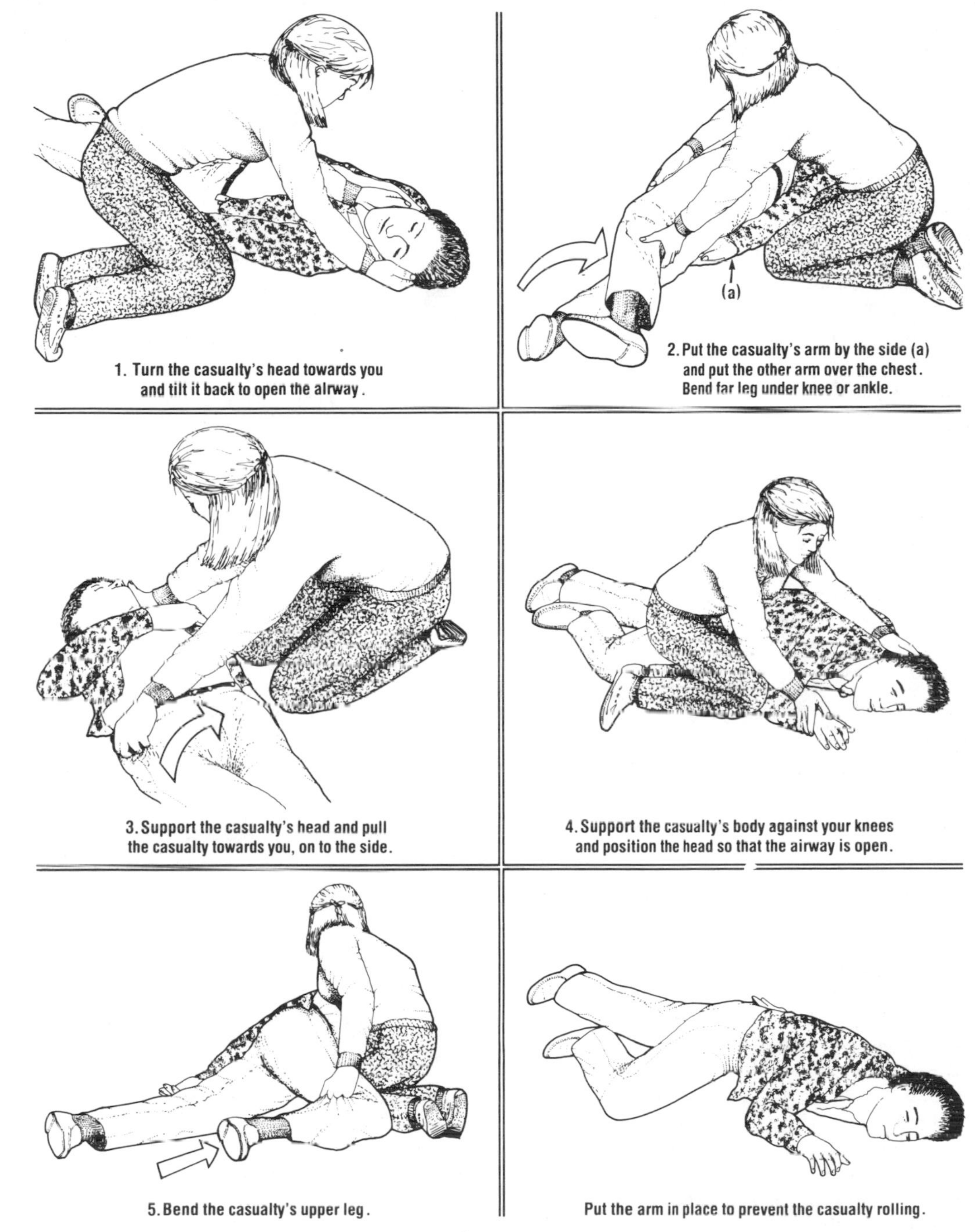

Fig. 25.2 The recovery position

There are times when the recovery position is not possible, such as if there are back or neck injuries, or if the casualty has fractures.

Bleeding

The normal adult has about 5½ litres (10 pints) of blood circulating around the body. Blood donors give 500 ml (a pint) of blood without suffering any ill effects, but losses of 1½ litres or more are dangerous and can lead to death. It is important for a first-aider to act quickly to prevent blood loss. The first way of doing this is by putting direct pressure on the wound (fig. 25.3) for 5 to 15 minutes. If there is a foreign body in the wound, the pressure needs to be applied beside it. Raising the injured limb will help reduce the flow of blood. If possible, firmly bandage the wound with a clean bandage.

Obviously, wounds vary a great deal. The deep, clean cuts caused by knives, razors, and so on bleed a great deal and take longer to clot than the jagged laceration wounds caused by barbed wire, or a scratch. However, lacerations and puncture wounds caused by nails, railings, etc. may lead to infection, and should be dealt with carefully.

There are three types of bleeding to consider:

1. Bleeding from *arteries* is bright red, as it is full of oxygen. The blood tends to spurt out of a wound as it is under pressure from the heart. This is the most serious type of bleeding and needs to be given priority.
2. Bleeding from *veins* is darker red, as it has less oxygen in it. The blood will not spurt out like blood from an artery unless a major vein is cut.
3. Most bleeding is from *capillaries*, which are the small blood vessels supplying the body with blood. The blood tends to come out slowly.

For a small wound, it is enough to run it under the tap to clean it, then pat it dry with cotton wool and put on a plaster.

Internal bleeding

Internal bleeding may be the result of an injury, or a bleeding stomach ulcer. Although the blood does not leave the body, it is just as dangerous as external bleeding, as the blood has left the circulatory system and the person will suffer the symptoms of blood loss. Apart from making the person as comfortable as possible and calling for an ambulance, there is nothing a first-aider can do to stop internal bleeding.

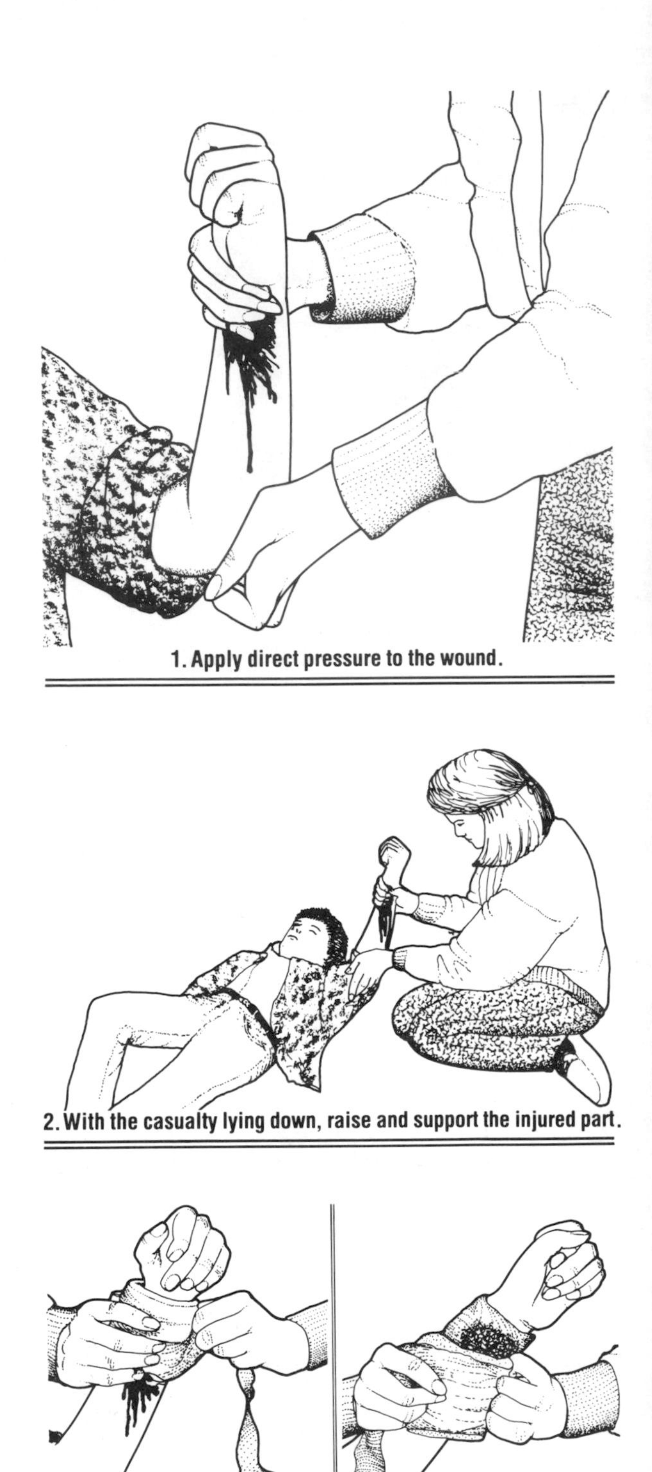

Fig. 25.3 Pressure and elevation to stop bleeding

Burns and scalds

Burns and scalds can be caused by heat, chemicals, or the sun:

- Dry burns are caused by heat found in fires, electrical appliances, and also by friction.
- Electrical burns are the result of electric shocks.
- The sun causes sunburn, which is very painful and may produce blistering.
- Scalds are caused by hot liquids, such as water or fat.
- Chemical burns are caused by corrosive acids or alkalis, found in household cleaners as well as in industry.

Burns are classified according to how severe they are:

1. When the outer layer of skin is burnt, it is called a *superficial burn*. These usually heal easily.
2. *Intermediate burns* are when blisters form, and there is then the added danger of infection.
3. *Deep burns* are very severe, yet may not be very painful because the nerves will have been damaged. The casualty of a severe burn needs to be taken away from the source of heat and treated.

Minor burns

A minor dry burn or scald should be held under cold running water for at least ten minutes for it to cool down. The area can then be dried and dressed with a clean, dry dressing. Points to remember:

- Do not break blisters, as they can become infected.
- Do not use sticking plaster on burns, as it will stick to them.
- Do not use soap, ointment, or grease on a burn.

Severe burns and scalds

With a severe burn or scald, any clothing that has stuck to the wound should be left alone, and a dry dressing should be put over the area until the casualty reaches hospital.

Chemical burns

Chemical burns should be run thoroughly under cold water before medical help is sought. It is important to find out which chemical caused the injury, as this may help the doctor treat the burn.

Chemicals in the eye should be treated by holding the eye under cold running water and then taking the casualty to hospital.

Sunburn

Sunburn is fairly common, and results in red and sore skin that feels hot to the touch. Mild sunburn can be treated with calamine lotion, but more severe cases should be taken to the doctor's or the hospital.

Electrical burns

An electric shock leads to burning. The appearance of the burn may seem slight, but there will be greater damage under the skin surface. The danger here is that the casualty may have stopped breathing, and the heart may have stopped beating.

The first thing the first-aider must do is to make sure that the casualty is away from the electrical current. The treatment is then the same as for other burns: put on a dry, sterile dressing and get medical help.

With all major burns and scalds, the casualty will probably also be suffering from shock, and will need to be treated for this as well as the injuries.

Choking

The symptoms of choking are the victim's gasping, clutching the throat, and becoming panicky. At this stage, the first-aider needs only to watch, as the victim may still be able to recover the foreign body by coughing. If the victim seems unable to dislodge the obstruction, then bend him or her over so that the head is lower than the lungs and slap the casualty firmly between the shoulder-blades up to four times. Check the mouth to see whether the obstruction has become dislodged. If the victim is still choking, repeat the procedure.

If the victim continues to choke, then the first-aider will have to apply abdominal thrusts. This technique can be dangerous and so needs to be demonstrated by a qualified first-aider. If you would like more information, contact your local British Red Cross Society or St John Ambulance Brigade.

Drowning

Drowning is an unfortunately frequent accident, and is often the result of carelessness. Although over two-thirds of people who drown are nonswimmers, good swimmers also drown for no apparent reason. Drowning is asphyxia (suffocation) by water, which can lead to death.

Certain safety precautions could reduce the number of people drowned each year, especially babies and young children, who need only a little water to drown in.

If the casualty is conscious when brought out of the water, then he or she should be treated for shock and be put on his or her back, with the legs and feet higher than the body, and kept warm and comfortable. As with all shock, nothing should be given to eat or drink, and a doctor should be called to check that the lungs are free of water.

If the casualty is unconscious, it is essential that air is put into the lungs as soon as possible by mouth-to-mouth resuscitation. If the first-aider is a competent swimmer, it is possible to do this in the water.

If there is a blockage in the throat, treat for choking by giving four sharp slaps between the shoulder-blades, and then four abdominal thrusts, before continuing with resuscitation.

If the water was cold, then the casualty may also need to be treated for hypothermia, as well as possible shock later. All drowning casualties should be seen by a doctor.

Poisoning

Although many cases of poisoning are attempted suicides, some are accidental poisonings caused by substances in everyday use.

It is best to take strict precautions to prevent accidental poisoning, especially where children are concerned, but it is also important to know how to treat cases of poisoning that do occur. Poisons can enter a person's body in four ways:

1. through the mouth, by the eating or drinking of poisonous substances;
2. by the breathing-in of poisonous fumes;
3. by absorption through the skin, such as pesticides or other chemicals;
4. when the skin is punctured, for example by an animal bite or insect sting.

Poisons reach the bloodstream and then cause various reactions. Some may lead to sickness, diarrhoea, and stomach pains; others may affect the central nervous system, stopping the heart and preventing breathing.

The general procedure and treatment for poisoning is:

- If the casualty is conscious, ask what has happened.
- If the casualty is unconscious and breathing, put him or her in the recovery position, but check the breathing regularly.
- If the heart stops, external chest compression should be started.

Points to remember:

- Never make a casualty vomit, because if the poison is corrosive it will damage the throat and mouth as it comes up.
- The first-aider should make sure that there are no traces of poison left around the casualty's mouth before starting mouth-to-mouth resuscitation.
- Look out for any containers near the casualty that may have held the poison. Take them with the casualty to hospital for diagnosis.

Poisonous fumes

If the casualty has breathed in poisonous fumes, then he or she must be put into fresh air. If the casualty is not breathing, or there is no heartbeat, then immediate treatment needs to be given.

If the casualty is unconscious in an enclosed space, the rescuers need to make sure that they can get the casualty out without becoming casualties themselves.

Absorbed poisons

Poisons absorbed through the skin may be quite mild, as in nettle rash, or more dangerous, as in certain pesticides. The best treatment is to cover the affected area with water and give the casualty plenty to drink. As always, any labels or containers should be kept to give information about the poison.

Punctured skin

Poisons reaching the bloodstream through punctured skin can be quite serious. In the UK, we are quite fortunate as we do not have many poisonous snakes and insects. Bee and wasp stings are fairly minor, unless the victim is allergic to them. The best treatment for minor stings is to put something cold on them, or apply soothing lotions, such as calamine.

In rarer cases, the victim may be very allergic to the content of the substance injected. This applies to medical injections of penicillin as well as wasp or bee stings. This condition is called *anaphylactic shock*, and the victim will have itchy and irritated skin, nettle rash, a swollen face and tongue, wheezing, weak pulse, and dizziness. He or she may also faint and go into a coma. The casualty should be immediately taken to hospital for an injection.

A person who has been stung in the mouth or throat needs to see a doctor as soon as possible, as there is a danger that the bite or sting may swell and block the air passage. Temporary help is to give a drink of cold water and put a cold compress on the throat.

Alcoholic poisoning

Alcoholic poisoning is reached when the person has drunk enough alcohol to make him or her incapable. The person may be vomiting, or unconscious. It is important to check that any vomit can escape without being inhaled.

Drug poisoning

Drug poisoning may be accidental, or it may be a suicide attempt. The drugs can be taken by mouth, injected into the bloodstream, or inhaled. The treatment is the same as for poisoning: don't make the casualty vomit, and check the breathing and heartbeat until medical help arrives.

Sudden emergencies

Sometimes it may be necessary to give first aid to someone who has been taken ill suddenly. A stroke victim may become unconscious – the first-aider then needs to check that the casualty's airway is kept open before the ambulance arrives. The victim of a fit, or convulsion, needs to be prevented from hurting himself or herself. There is no need to use force to prevent the person from moving, or to put anything in the mouth to stop him or her from biting the tongue.

The important thing to remember when an accident or emergency happens is that speed is essential. We should all have a basic understanding of first-aid techniques and know how to use the contents of a first-aid kit. We should also know how to get medical help.

Asphyxia

Asphyxia, or suffocation, occurs when there is not enough oxygen reaching the body. This may be due to choking, or drowning – which have already been covered – or any of the following:

- pressure on the chest, such as that caused by being crushed in a crowd, or by a fall of sand or earth
- suffocation by pillows, polythene bags, etc.
- hanging or strangulation
- injured lungs
- injured chest
- airway obstructed, e.g. by inhaled vomit or by a swelling caused by an insect sting

People can also suffocate from lack of oxygen in the air they breathe – in smoke-filled buildings, this is as much of a danger as fire itself.

High or low altitude can also lead to suffocation. At high altitude, there is too little oxygen in the air, and at low altitude there is a great deal of pressure.

If the brain or the nerves that control breathing are affected by electrocution, poisoning, or paralysis, then the person will be unable to breathe unaided and will suffocate.

The treatment depends on the cause of asphyxiation, and the cause should be removed before starting mouth-to-mouth resuscitation. For example, if a child has a polythene bag over the head, this will have to be removed before the child can be resuscitated. The person buried under a fall of sand will need to be dug out before first aid can be given.

All asphyxia casualties should be taken to hospital for a check-up.

Assignments on basic first aid

For assignments marked with an asterisk (*), it is necessary to have a first-aid demonstration carried out by a trained instructor. For advice, contact your local British Red Cross Society or St John Ambulance Brigade.

1. The results of a MORI poll carried out for the St John Ambulance Brigade showed these disturbing facts:
    - Six out of ten people suggested potentially dangerous treatment for heart-attack victims.
    - Six out of ten did not know how to deal with a nose-bleed.

- Although more men had first-aid training than women, more women knew the correct first-aid treatment.

Do you think that more should be done about training the public in first aid at school and at work?

2. If a wound – particularly one from a rusty nail – is not carefully cleaned, tetanus infection can develop, and this can cause death. Find out the symptoms, causes, treatment, and prevention of tetanus.
3. What are the symptoms of severe blood loss?

*4. Resuscitation may save a person's life, so it is important that we learn to do it properly. It is useful to practise on one of the models available.

- How do you check whether a person is breathing?
- How do you open the airway before giving mouth-to-mouth resuscitation?
- Give a practical demonstration of mouth-to-mouth resuscitation for an adult.
- Give a practical demonstration of mouth-to-mouth resuscitation for a child.

*5. External chest compression/heart massage must be done properly in order to save a life with the least physical damage to the casualty. Demonstrate the process.

*6. Practise putting a casualty into the recovery position, following the guidelines given in the first-aid manual or by the first-aid instructor.

*7. Clinical shock is very dangerous, so it is important to know how to recognise and treat it. Check that you understand:

- the signs of clinical shock
- the causes
- how to treat a shocked casualty

8. Before treating a casualty of electric shock, it is necessary to separate the casualty from the electrical current. Think of ways to do this.
9. Find out the causes, symptoms, and treatment of: (a) hypothermia, (b) heatstroke.
10. Make a list of the types of burn that could be caused in these situations:

- sliding down a rope too fast
- getting a cake out of the oven without oven-gloves
- working in a chemical factory without gloves
- splashes from a chip-pan on to hands and arms
- touching a hot car exhaust pipe

How would you deal with each of these burns or scalds?

11. Make some safety guidelines to try to prevent people drowning:

- in boats
- swimming in the sea
- babies and young children
- through the hazards of ice

12. Write a set of precautions to prevent some of the accidental poisonings that happen every week:

- at work
- on the farm
- at home: household cleaners, do-it-yourself, cosmetics
- in the garden: poisonous plants, weed-killers

13. Food poisoning may be caused by the bacteria staphylococci and salmonella. Which is the most common cause of food poisoning?
    Find out:

- how the bacteria are passed on to the person
- the symptoms
- the treatment
- ways of preventing infection

14. What are the symptoms of these forms of poisoning:

- carbon monoxide
- pesticides (parathion and malathion)
- urticaria (nettle rash)

15. How should you deal with a poisonous snake bite?
16. Find out about rabies:

- What is it?
- How is it passed on to man?
- What are the symptoms?
- Is there any treatment?
- How have we managed to prevent it from becoming established in the UK?

*17. Dressings and bandages come in many shapes and sizes, and the best way to learn their uses is by a practical demonstration. Practise using these on each other:

sterile dressings	triangular bandage
gauze dressings	ring pads
plasters	reef knots
adhesive strapping	slings
roller bandage	tubular gauze

Cover these injury situations:

- Bandage a wound with a foreign body.
- Bandage a head injury.
- Bandage a deep cut.

18. Find out the causes and symptoms of minor and major epilepsy.
19. Find out the causes and symptoms of diabetes. Include hypoglycaemia and hyperglycaemia in your notes, and explain the difference between them.
20. Find out the treatment for removing foreign bodies from the eye.
21. It is important to get proper first-aid training, as we never know when we may have to act quickly in an accident or emergency. Find out the various agencies running first-aid courses in your area.

    - What is the length of the course?
    - What is the cost of the course?
    - Is there a certificated examination at the end of the course?
    - Does the course have to be up-dated at intervals?

22. The ambulance service deals with most emergencies and accidents. Find out exactly how to call an ambulance and what information you need to give.

    There are other emergency services available: find out any information you can about them. Here are some ideas:

    - air–sea rescue
    - Royal National Lifeboat Institute
    - poison centre
    - mountain rescue

23. Dealing with a house on fire may never happen to you. However, knowing the right thing to do can mean the difference between life and death. If possible, arrange either a visit to or a talk from your local fire brigade. It is very useful and will explain exactly how we can help prevent fire from spreading. If a visit or talk is not possible, research into fires and find out exactly what should be done.

Index